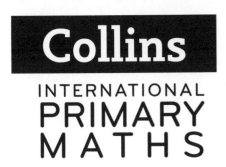

Collins

INTERNATIONAL PRIMARY MATHS

Workbook 2

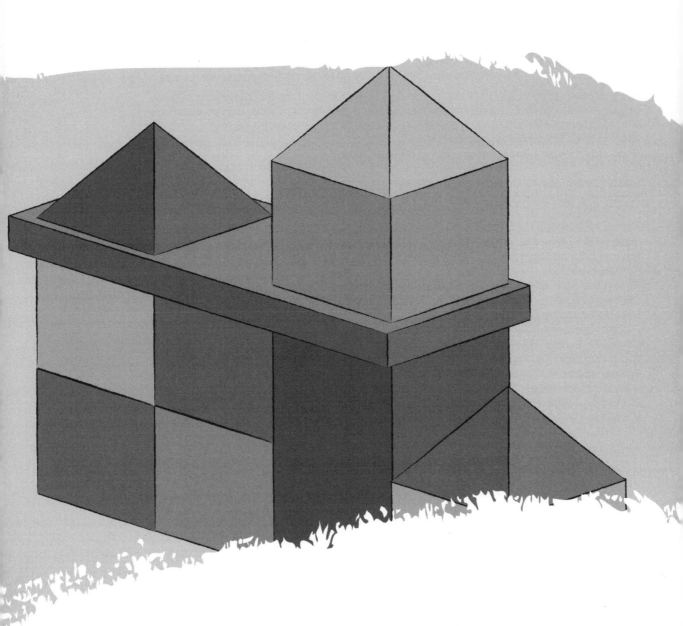

William Collins' dream of knowledge for all began with the publication of his first book in 1819. A self-educated mill worker, he not only enriched millions of lives, but also founded a flourishing publishing house. Today, staying true to this spirit, Collins books are packed with inspiration, innovation and practical expertise. They place you at the centre of a world of possibility and give you exactly what you need to explore it.

Collins. Freedom to teach.

Published by Collins
An imprint of HarperCollins*Publishers*
The News Building
1 London Bridge Street
London
SE1 9GF

HarperCollins*Publishers*
Macken House,
39/40 Mayor Street Upper,
Dublin 1,
D01 C9W8, Ireland

Browse the complete Collins catalogue at
www.collins.co.uk

© HarperCollins*Publishers* Limited 2021

10 9 8 7 6

ISBN 978-0-00-836946-0

British Library Cataloguing-in-Publication Data
A catalogue record for this publication is available from the British Library.

Author: Lisa Jarmin
Series editor: Peter Clarke
Publisher: Elaine Higgleton
Product developer: Holly Woolnough
Project manager: Mike Harman (Life Lines Editorial Services)
Development editor: Joan Miller
Copyeditor: Catherine Dakin
Proofreader: Tanya Solomons
Cover designer: Gordon MacGilp
Cover illustrator: Ann Paganuzzi
Typesetter: Ken Vail Graphic Design
Illustrators: Ann Paganuzzi, Ken Vail Graphic Design and QBS Learning
Production controller: Lyndsey Rogers
Printed and bound in India by Replika Press Pvt. Ltd.

With thanks to the following teachers and schools for reviewing materials in development: Antara Banerjee, Calcutta International School; Hawar International School; Melissa Brobst, International School of Budapest; Rafaella Alexandrou, Pascal Primary Lefkosia; Maria Biglikoudi, Georgia Keravnou, Sotiria Leonidou and Niki Tzorzis, Pascal Primary School Lemessos; Taman Rama Intercultural School, Bali.

Contents

Number

Geometry and Measure

Statistics and Probability

How to use this book

This book is used during the part of a lesson when you practise the mathematical ideas you have just been taught.

- An **objective** explains what you should know, or be able to do, by the end of the lesson.

You will need

- Shows the things you need to use to answer some of the questions.

There is a page of practice questions for each lesson, with three different types of questions:

1 Some question numbers are written on a **circle**. These questions may be **easier**. They may also practise ideas you have learned before. These questions will help you answer the rest of the questions on the page.

2 Some question numbers are written on a **triangle**. These questions help you better understand the ideas you have just been taught.

3 Some question numbers are written on a **square**. These are a little **harder** and make you think.

You won't always have to answer all the questions on the page. Your teacher will tell you which questions to answer.

HINT

Draw a ring around the question numbers your teacher tells you to answer.

Date: _____

At the bottom of the page there is room to write the date you completed the work. If it took you longer than 1 day, write all of the dates you worked on the page.

Self-assessment

Once you've answered the questions, think about how easy or hard you find the ideas. Draw a ring around the face that describes you best.

😊 I can do this.

😐 I'm getting there.

☹ I need some help.

Lesson 1: **Counting in ones**

• Count objects in 1s

1 Cross out each sweet as you count it.

How many are there? ☐

2 Cross out each marble as you count it.

How many are there? ☐

3 Cross out each star as you count it.

How many are there? ☐

4 Draw 39 circles in the box. Count them to check that you have drawn the correct amount.

Date: _____

Number

Lesson 2: **Recognising patterns**

- Recognise up to 10 objects in unfamiliar patterns without counting

1 Without counting, write how many dots are in each ten frame.

a [ ]

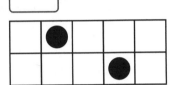

b [ ]

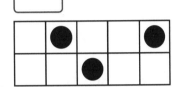

2 Without counting, write how many dots are in each ten frame.

a [ ]

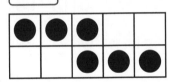

b [ ]

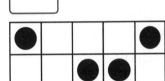

c [ ]

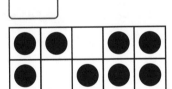

d [ ]

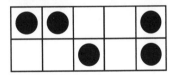

3 Without counting, do you think these ten frames contain the same amount of dots or different amounts?

_____

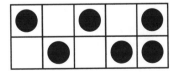

Date: _____

7

Number

Lesson 3: **Counting in fives**

- Count on and back in 5s
- Count objects by making groups of 5

1 Write the next number in each sequence. Use the 100 square to help you.

1	2	3	4	5	6	7	8	9	10
11	12	13	14	15	16	17	18	19	20
21	22	23	24	25	26	27	28	29	30
31	32	33	34	35	36	37	38	39	40
41	42	43	44	45	46	47	48	49	50
51	52	53	54	55	56	57	58	59	60
61	62	63	64	65	66	67	68	69	70
71	72	73	74	75	76	77	78	79	80
81	82	83	84	85	86	87	88	89	90
91	92	93	94	95	96	97	98	99	100

a 5, 10, 15, 20, ☐

b 40, 45, 50, 55, ☐

c 65, 70, 75, 80, ☐

2 Draw a ring around each group of 5 marbles. Then count the marbles in 5s.

☐

3 Draw a ring around each group of 5 counters. Then count the counters in 5s. Add on any leftover counters at the end. ☐

Date: _____

Number

Lesson 4: **Counting in tens**

- Count on and back in 10s
- Count objects by making groups of 10

1 Write the next number in each sequence. Use the 100 square to help you.

1	2	3	4	5	6	7	8	9	10
11	12	13	14	15	16	17	18	19	20
21	22	23	24	25	26	27	28	29	30
31	32	33	34	35	36	37	38	39	40
41	42	43	44	45	46	47	48	49	50
51	52	53	54	55	56	57	58	59	60
61	62	63	64	65	66	67	68	69	70
71	72	73	74	75	76	77	78	79	80
81	82	83	84	85	86	87	88	89	90
91	92	93	94	95	96	97	98	99	100

a 10, 20, 30, 40, ☐

b 40, 50, 60, 70, ☐

c 60, 70, 80, 90, ☐

2 Draw a ring around each group of 10 jewels. Then count the jewels in 10s.

☐

3 Draw a ring around each group of 10 pearls. Then count the pearls in 10s. Add on any leftover pearls at the end.

☐

Date: _____

9

Number

Lesson 1: **Counting in twos**

- Count on and back in 2s
- Count objects in groups of 2

1 Write the next number in each pattern. Use the 100 square to help.

1	2	3	4	5	6	7	8	9	10
11	12	13	14	15	16	17	18	19	20
21	22	23	24	25	26	27	28	29	30
31	32	33	34	35	36	37	38	39	40
41	42	43	44	45	46	47	48	49	50
51	52	53	54	55	56	57	58	59	60
61	62	63	64	65	66	67	68	69	70
71	72	73	74	75	76	77	78	79	80
81	82	83	84	85	86	87	88	89	90
91	92	93	94	95	96	97	98	99	100

a 2, 4, 6, 8, 10, ☐

b 36, 38, 40, 42, ☐

c 72, 74, 76, 78, ☐

2 Draw a ring around each group of 2 bears. Then count the bears in 2s.

☐

3 Draw a ring around each group of 2 balls. Then count the balls in 2s. If there is one left over, add it on. ☐

Date: _____

😊 😐 ☹

Number

Lesson 2: **Even and odd numbers**

• Recognise even and odd numbers to 100

You will need
• coloured pencil

1 Colour the even numbers.

| 1 | 2 | 3 | 4 | 5 | 6 | 7 | 8 | 9 | 10 | 11 | 12 | 13 | 14 | 15 | 16 | 17 | 18 | 19 | 20 |

2 Draw a line to join each boat to the odd or even bank.

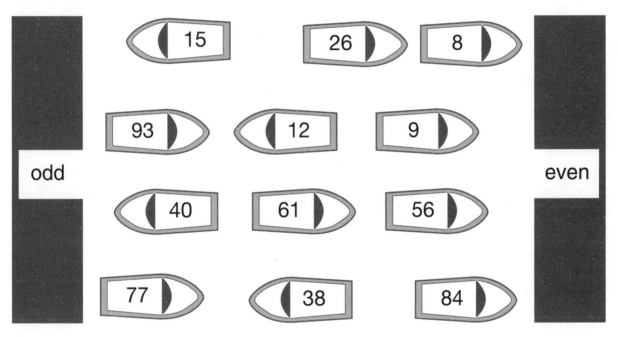

3 Colour the number that should **not** be in this group.

⟨17⟩ ⟨55⟩ ⟨21⟩ ⟨36⟩ ⟨19⟩

4 a Colour the numbers that will share into 2 groups equally.

⟨78⟩ ⟨52⟩ ⟨33⟩ ⟨19⟩ ⟨1⟩ ⟨66⟩ ⟨80⟩ ⟨95⟩ ⟨61⟩

b How did you know the numbers would share equally into 2 groups?

_____

_____

Date: _____

11

Number

Lesson 3: **Counting on and back**

• Count on and back in steps of 1, 2, 5 and 10

1 Continue the number patterns.

a

2 4 6 8 10

b

10 20 30 40 50

2 Continue the number patterns.

a

10 15 20 25

b

16 14 12 10 8

3 Draw a ring around the answers in this sentence. The pattern on the stars is counting | on | back | in: | 2s | 5s | 10s |.

30 25 20 15 10

4 Continue the number patterns.

a | 34 | 44 | 54 | 64 | | | |

The pattern is counting on in: ☐

b | 12 | 17 | 22 | 27 | 32 | | | | |

The pattern is counting on in: ☐

Date: _____

Lesson 4: **Estimating**

- Estimate how many objects in a set of 20–100

1 Draw a ring around the estimate that you think is closer to the number of planets.

20   60

2 Draw a ring around the estimate that you think is closest to the number of stars.

20   50   70   100

3 Draw a ring around the estimate that you think is closest to the number of rockets.

20   50   70   100

4 Estimate how many stars there are to the nearest 10. [ ]

5 Look at the stars in **2**. Draw rings around groups of stars to find out how many stars there are. [ ]

Date: _____

Number

Lesson 1: **Counting to 100**

- Count on in 1s from 0 to 100
- Count back in 1s from 100 to 0

You will need
- Slide 1 or the Student's Book

1 Draw a line from 20 and count **on**. Use the game board if you need to.

20   24   28

21   23   25   27   29   31

22   26   30

2 Draw a line from 48 and count **on**.

49   50   54   55   57

48   51   52   53   56

3 Draw a line from 83 and count **back**.

83   82   79   78   75

81   80   77   76   74

4 Count on from 61. What numbers belong in the last four stars?

61 62 63 64 65 66 67 68

Date: _____

Lesson 2: **Reading numbers to 100**

• Read numbers from 0 to 100

Number

1 Say the number on each fish with your partner.

 6 15 19  20 21 27

2 Say the number on each shell. Then draw a line to match each fish and shell.

 86 52  60 22 12 99 35 41

 22 52 12 86 41 60 99 35

3 Say the number on each jellyfish. Then draw a line to match to the number that comes next when you count forwards.

 50  39  22

99 51 23 11 78 40

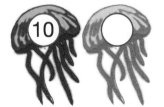 10  98  77

Lesson 3: **Writing numbers to 100**

Number

• Write numbers from 0 to 100

1 Fill in the missing numbers.

a

21	22		24			27		29	30

b

51		53		55	56		58		

2 Fill in the missing numbers.

a

	22	23	24	25		27	28	29	30
	32	33	34		36	37	38	39	
	42	43	44	45	46		48	49	50

b

61			64	65		67	68		70
71		73		75	76	77		79	80
	82	83	84		86		88	89	
91	92		94	95		97	98		

3 Fill in the missing numbers.

	56	57	58		
65	66		68	69	70
75		77	78	79	80
	86	87		89	
95	96		98		100

Date: _____

Lesson 4: **Reading and writing number names to 100**

• Read and write number names from 0 to 100

You will need
• Slide 1

1 Draw lines to match the numbers to their number names.

23 •   • thirty-three

41 •   • twenty-three

80 •   • eighty

99 •   • sixty-two

33 •   • forty-one

62 •   • ninety-nine

2 Write the matching number names.

| 50 | _____ | 24 | _____ |
| 97 | _____ | 66 | _____ |

3 Write the matching numbers.

| thirty | ☐ | eighty-two | ☐ |
| forty-one | ☐ | seventy-nine | ☐ |

4 Write the number names to match.

How many learners are in your class? _____

What is the page number for this lesson? _____

Date: _____   ☺ ☺ ☹

Lesson 1: **The link between addition and subtraction**

- Understand that addition is the opposite of subtraction
- Record matching addition and subtraction number sentences

1 Write a matching subtraction.

$3 + 1 = 4$

$4 - \boxed{} = \boxed{}$

2 Write a matching addition.

$3 - 2 = 1$

$\boxed{} + \boxed{} = 3$

3 Work out the answer. Then write one addition and two subtractions to match.

$5 + 3 = \boxed{}$

$\boxed{} + \boxed{} = \boxed{}$

$\boxed{} - \boxed{} = \boxed{}$

$\boxed{} - \boxed{} = \boxed{}$

4 Work out the answer. Then write one subtraction and two additions to match.

$9 - 6 = \boxed{}$

$\boxed{} - \boxed{} = \boxed{}$

$\boxed{} + \boxed{} = \boxed{}$

$\boxed{} + \boxed{} = \boxed{}$

5 Work out the answer. Then write one addition and two subtractions to match.

$34 + 5 = \boxed{}$

$\boxed{} + \boxed{} = \boxed{}$

$\boxed{} - \boxed{} = \boxed{}$

$\boxed{} - \boxed{} = \boxed{}$

6 Work out the answer. Then write one subtraction and two additions to match.

$57 - 3 = \boxed{}$

$\boxed{} - \boxed{} = \boxed{}$

$\boxed{} + \boxed{} = \boxed{}$

$\boxed{} + \boxed{} = \boxed{}$

Date: _____

Lesson 2: **Making 20**

• Know pairs of numbers that total 20

1 Draw extra dots to make 20.

a

20	
●●●●●●●●●● ●●●●●●●●●●	

b

20	
●●●●●●●● ●●●●●●●	

2 Complete the diagrams to show pairs of numbers that total 20. Then complete the number sentences.

a

20	
7	

7 + ☐ = 20

20 − ☐ = 7

☐ + 7 = 20

20 − 7 = ☐

b

20	
1	

1 + ☐ = 20

20 − ☐ = 1

☐ + 1 = 20

20 − 1 = ☐

c

20	
15	

15 + ☐ = 20

20 − ☐ = 15

☐ + 15 = 20

20 − 15 = ☐

d

20	
12	

12 + ☐ = 20

20 − ☐ = 12

☐ + 12 = 20

20 − 12 = ☐

3 Complete the diagram and the number sentences.

20	

☐ + ☐ = 20   ☐ + ☐ = 20

20 − ☐ = ☐   20 − ☐ = ☐

Date: _____

Number

Lesson 3: **Adding and subtracting tens**

• Add and subtract multiples of 10

1 Draw a ring around the pair of tens that equal 30.

20 + 50   10 + 20   70 + 10

2 Draw a ring around the pair of tens that equal 40.

10 + 10   90 + 10   30 + 10

3 Write an addition fact and a subtraction fact for these tens.

a

40	
20	20

☐ + ☐ = ☐

☐ − ☐ = ☐

b

70	
60	10

☐ + ☐ = ☐

☐ − ☐ = ☐

c

90	
60	30

☐ + ☐ = ☐

☐ − ☐ = ☐

d

50	
20	30

☐ + ☐ = ☐

☐ − ☐ = ☐

4 Complete each number sentence.

a 50 + 20 = ☐   **b** 80 − 20 = ☐   **c** 20 + 40 = ☐

d 90 − 50 = ☐   **e** 30 + ☐ = 70   **f** 60 − ☐ = 30

5 Write three different pairs of multiples of 10 that total 80.

☐ + ☐ = 80   ☐ + ☐ = 80   ☐ + ☐ = 80

Date: _____

Lesson 4: **Adding more than two numbers**

• Add more than two small numbers together

1 Use the number line to help you answer these questions.

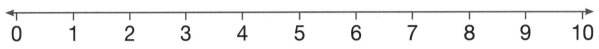

a $3 + 2 + 1 =$ ☐ **b** $5 + 2 + 2 =$ ☐

c $2 + 1 + 1 =$ ☐ **d** $4 + 2 + 1 =$ ☐

2 Order the numbers, then add them together. Estimate first.

a $3 + 4 + 1 + 2 =$ ☐ Estimate: ☐

b $2 + 2 + 1 + 3 =$ ☐ Estimate: ☐

c $1 + 2 + 3 + 3 =$ ☐ Estimate: ☐

3 Write the number sentence, then add the numbers together.

I bought 3 marbles, 1 teddy, 3 stickers, 2 cars and 1 ball.
How many things did I buy altogether?

☐ + ☐ + ☐ + ☐ + ☐ = ☐

Date: _____

Lesson 1: **Adding 2-digit numbers and ones**

• Add a 1-digit number to a 2-digit number

1 Add the ones together. Then add them to the tens.

a 64 + 2 = ☐

b 55 + 3 = ☐

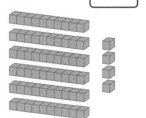

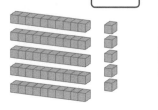

2 Use the number line to count on from the greater number. Estimate first.

a 82 + 5 = ☐   Estimate: ☐

b 25 + 4 = ☐   Estimate: ☐

c 8 + 71 = ☐   Estimate: ☐

3 Solve these by partitioning the numbers. Estimate first.

a 43 + 5 = ☐

Estimate: ☐

b 62 + 7 = ☐

Estimate: ☐

Date: _____

Number

Lesson 2: **Adding 2-digit numbers and tens**

• Add tens to a 2-digit number

1 Add the tens together. Then count on the ones.

a 25 + 20 = ☐

b 41 + 30 = ☐

2 Use the number line to count on in tens.

a 39 + 30 = ☐

b 47 + 50 = ☐

c 23 + 40 = ☐

d 68 + 10 = ☐

3 Solve these by partitioning the numbers.

a 40 + 49 = ☐

b 60 + 25 = ☐

Date: _____

Number

Lesson 3: **Adding 2-digit numbers (1)**

• Add pairs of 2-digit numbers

1 Add the tens first. Then add on the ones.

a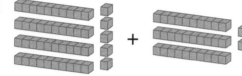

b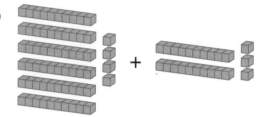

☐ + ☐ = ☐   ☐ + ☐ = ☐

2 Use the number line to add on the tens first, then the ones. Estimate first.

a 34 + 25 = ☐   Estimate: ☐

34

b 77 + 12 = ☐   Estimate: ☐

77

c 53 + 32 = ☐   Estimate: ☐

53

3 Use the number line to solve the problem. Estimate first.

55 people are sitting on a plane. 33 seats are empty.

How many seats are on the plane? = ☐

Estimate: ☐

Date: _____

Number

Lesson 4: **Adding 2-digit numbers (2)**

- Add pairs of 2-digit numbers

1 Complete each addition.

a
```
    5   4
+   1   2
        6
   ☐   ☐
   ☐   ☐
```

b
```
    7   6
+   2   1
       ☐
    9   0
   ☐   ☐
```

c
```
    4   3
+   2   5
        8
   ☐   ☐
   ☐   ☐
```

2 Work out the answers using the expanded written method.

a
```
    4   4
+   3   2
       ☐
   ☐   ☐
   ☐   ☐
```

b
```
    6   5
+   3   2
       ☐
   ☐   ☐
   ☐   ☐
```

c
```
    8   3
+   1   6
       ☐
   ☐   ☐
   ☐   ☐
```

3 Work out the answers using the formal written method.

a 55 + 34 = ☐

```
    ☐   ☐
+   ☐   ☐
   ☐   ☐
```

b 43 + 36 = ☐

```
    ☐   ☐
+   ☐   ☐
   ☐   ☐
```

c 25 + 53 = ☐

```
    ☐   ☐
+   ☐   ☐
   ☐   ☐
```

Date: _____

Number

Lesson 1: **Subtracting 2-digit numbers and ones**

• Subtract a 1-digit number from a 2-digit number

1 Cross out the ones. Then count how many are left.

a 59 – 4 = ☐   **b** 37 – 6 = ☐

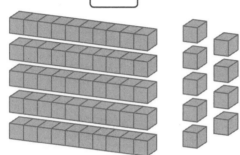

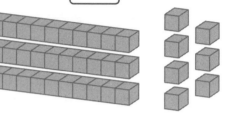

2 Use the number line to count back. Estimate first.

a 26 – 2 = ☐   Estimate: ☐

b 55 – 4 = ☐   Estimate: ☐

c 69 – 5 = ☐   Estimate: ☐

3 Count back to solve these subtractions. Estimate first.

a 96 – 4 = ☐   Estimate: ☐

b 78 – 7 = ☐   Estimate: ☐

Date: _____

Lesson 2: **Subtracting 2-digit numbers and tens**

• Subtract tens from a 2-digit number

1 Cross out the tens. Then count the tens and ones that are left.

a $52 - 30 = $ ☐

b $46 - 20 = $ ☐

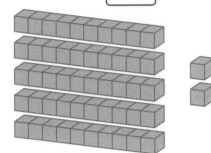

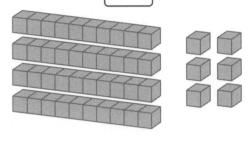

2 Use the number line to count back in tens.

a $93 - 40 = $ ☐

93

b $79 - 50 = $ ☐

79

c $23 - 10 = $ ☐

23

d $56 - 30 = $ ☐

56

3 Count back in tens to solve the subtractions.

a $85 - 40 = $ ☐   **b** $61 - 40 = $ ☐   **c** $77 - 30 = $ ☐

Date: _____

Lesson 3: **Subtracting 2-digit numbers (1)**

Number

• Subtract pairs of 2-digit numbers

1 Cross out the tens and ones to solve the subtractions.

a 58 – 25 = [   ]

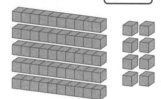

b 76 – 34 = [   ]

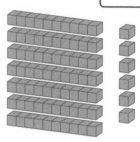

2 Use the place value charts to solve the subtractions.
Estimate first.

a 53 – 22 = [   ]   **b** 86 – 34 = [   ]   **c** 77 – 41 = [   ]

Estimate: [   ]   Estimate: [   ]   Estimate: [   ]

10s	1s
❘ ❘ ❘ ❘ ❘	○○○

10s	1s
❘ ❘ ❘ ❘ ❘	○○○○
❘ ❘ ❘	○○

10s	1s
❘ ❘ ❘ ❘ ❘	○○○○
❘ ❘	○○○

3 Fill in the place value charts to solve the subtractions.
Estimate first.

a 68 – 34 = [   ]   Estimate: [   ]

10s	1s

b 89 – 62 = [   ]   Estimate: [   ]

10s	1s

Date: _____

Lesson 4: **Subtracting 2-digit numbers (2)**

Number

• Subtract one 2-digit number from another

1 Complete each subtraction.

a 68 – 25 = ☐

60	8
– 20	5
☐	☐

b 49 – 13 = ☐

40	9
– 10	3
☐	☐

c 57 – 32 = ☐

50	7
– 30	2
☐	☐

2 Work out the answers using the expanded written method. Estimate first.

a 75 – 42 = ☐

Estimate: ☐

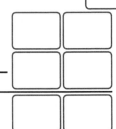

b 68 – 17 = ☐

Estimate: ☐

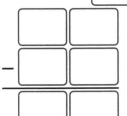

c 36 – 22 = ☐

Estimate: ☐

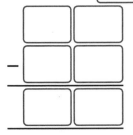

3 Work out the answers using the formal written method. Estimate first.

a 59 – 42 = ☐

Estimate: ☐

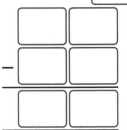

b 78 – 53 = ☐

Estimate: ☐

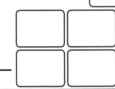

c 87 – 66 = ☐

Estimate: ☐

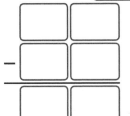

Date: _____

Number

Lesson 1: **Multiplication as repeated addition (1)**

- Count objects in groups to solve multiplication problems with repeated addition

1 There are ☐ bags. There are ☐ apples in each bag.

Add: ☐ + ☐ + ☐ = ☐

Multiply: ☐ × ☐ = ☐

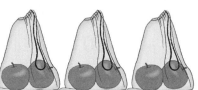

2 Count the fruit in their groups to solve each multiplication.

a $2 \times 5 = $ ☐   🍒 + 🍒 + 🍒 + 🍒 + 🍒 = ☐

b $5 \times 3 = $ ☐   🍌 + 🍌 + 🍌 = ☐

c $10 \times 2 = $ ☐   🍇 + 🍇 = ☐

d $2 \times 6 = $ ☐   🍓 + 🍓 + 🍓 + 🍓 + 🍓 + 🍓 = ☐

3 Fill in the multiplication to match the addition.

a [🔵🔵🔵🔵🔵] + [🔵🔵🔵🔵🔵] + [🔵🔵🔵🔵🔵] + [🔵🔵🔵🔵🔵] = ☐

☐ × ☐ = ☐

b [🔵🔵] + [🔵🔵] + [🔵🔵] + [🔵🔵] + [🔵🔵] + [🔵🔵] + [🔵🔵] = ☐

☐ × ☐ = ☐

c [🔵🔵🔵🔵] + [🔵🔵🔵🔵] + [🔵🔵🔵🔵] = ☐

☐ × ☐ = ☐

Date: _____

Lesson 2: **Multiplication as repeated addition (2)**

> • Use a diagram to solve multiplication problems with repeated addition

1 Count dots in 2s or 10s to solve the multiplications.

a $2 \times 5 =$ ☐ ⬤⬤ | ⬤⬤ | ⬤⬤ | ⬤⬤ | ⬤⬤

b $10 \times 3 =$ ☐ ⬤⬤⬤⬤⬤ ⬤⬤⬤⬤⬤ | ⬤⬤⬤⬤⬤ ⬤⬤⬤⬤⬤ | ⬤⬤⬤⬤⬤ ⬤⬤⬤⬤⬤

2 Draw crosses in the diagrams to solve the multiplications.

a $5 \times 6 =$ ☐ | X X X X X | | | | |

b $2 \times 9 =$ ☐ | X X | | | | | | | |

c $10 \times 4 =$ ☐ | X X X X X X X X X X | | | |

d $5 \times 3 =$ ☐ | X X X X X | | |

3 Draw crosses in the diagrams to solve the multiplications.

a $5 \times 9 =$ ☐

b $2 \times 11 =$ ☐

Date: _____

Number

Lesson 3: **Multiplication using a number line**

- Use a number line to solve multiplication problems with repeated addition

1 Count along the number line in 2s to find each answer.

a $2 \times 2 =$ ☐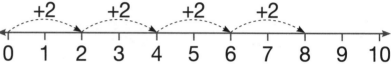

b $2 \times 4 =$ ☐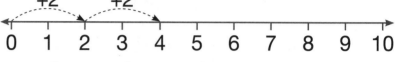

2 Draw the correct number of jumps to match each multiplication. Then write the answer.

a $2 \times 5 =$ ☐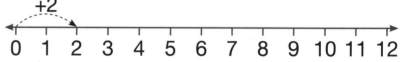

b $2 \times 3 =$ ☐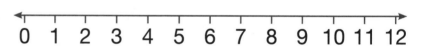

c $2 \times 6 =$ ☐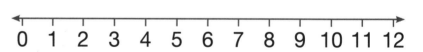

d $3 \times 4 =$ ☐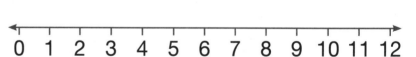

3 Draw the correct number of jumps to match each multiplication. Then write the answer.

a $6 \times 3 =$ ☐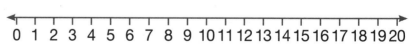

b $4 \times 4 =$ ☐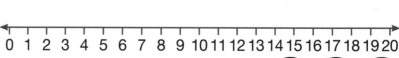

Date: _____

Lesson 4: **Multiplication using a 100 square**

- Use a 100 square to solve multiplication problems with repeated addition

1 Count along in 5s four times on the number grid to solve the multiplication.

1	2	3	4	5	6	7	8	9	10
11	12	13	14	15	16	17	18	19	20
21	22	23	24	25	26	27	28	29	30

$5 \times 4 =$ ☐

2 Use the 100 square to help you to solve the multiplications.

1	2	3	4	5	6	7	8	9	10
11	12	13	14	15	16	17	18	19	20
21	22	23	24	25	26	27	28	29	30
31	32	33	34	35	36	37	38	39	40
41	42	43	44	45	46	47	48	49	50
51	52	53	54	55	56	57	58	59	60
61	62	63	64	65	66	67	68	69	70
71	72	73	74	75	76	77	78	79	80
81	82	83	84	85	86	87	88	89	90
91	92	93	94	95	96	97	98	99	100

a $10 \times 6 =$ ☐

b $2 \times 9 =$ ☐

c $5 \times 10 =$ ☐

d $2 \times 7 =$ ☐

e $10 \times 9 =$ ☐

f $5 \times 6 =$ ☐

3 Use the 100 square in **2** to help you to solve the multiplications.

a $4 \times 3 =$ ☐   **b** $3 \times 6 =$ ☐

c $4 \times 7 =$ ☐   **d** $3 \times 8 =$ ☐

e $4 \times 9 =$ ☐   **f** $3 \times 10 =$ ☐

Date: _____ ☺ ☺ ☹

Lesson 1: **Multiplication as an array (1)**

• Understand multiplication as an array

1 Use the arrays to solve the multiplications.

a

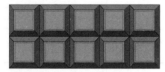

2 × 5 = ☐

b

2 × 4 = ☐

2 Write and solve a number sentence for each array.

a

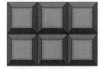

☐ × ☐ = ☐

b

☐ × ☐ = ☐

c

☐ × ☐ = ☐

d

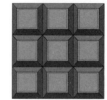

☐ × ☐ = ☐

3 Write and solve one number sentence for each array. Then turn your book on its side and look at the arrays sideways. Write a different number sentence to match each array.

a

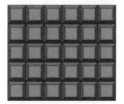

☐ × ☐ = ☐

☐ × ☐ = ☐

b

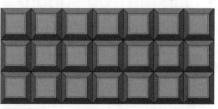

☐ × ☐ = ☐

☐ × ☐ = ☐

Date: _____

34

Lesson 2: **Multiplication as an array (2)**

• Draw arrays to solve multiplication problems

1 Complete the arrays to match each number sentence.

a $3 \times 5 = \boxed{}$

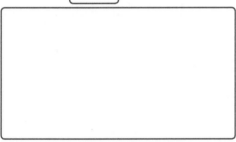

b $2 \times 6 = \boxed{}$

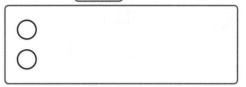

2 Draw an array to match each number sentence.

a $3 \times 6 = \boxed{}$

b $5 \times 5 = \boxed{}$

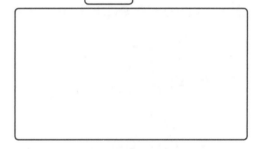

c $2 \times 9 = \boxed{}$

d $2 \times 4 = \boxed{}$

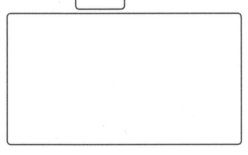

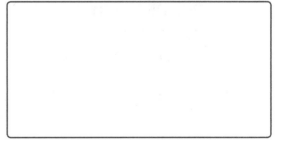

3 Write a multiplication number sentence using any two numbers from 3 to 7, then draw an array to solve it.

$\boxed{} \times \boxed{} = \boxed{}$

Date:

Number

Number

Lesson 3: **The equals sign**

- Understand that an array can show two multiplications with the same answer
- Understand that the facts on either side of the equals sign have the same value

1 Complete the equal number statement to match each array.

a

$2 \times 5 = \boxed{} \times \boxed{}$

b

$4 \times 3 = \boxed{} \times \boxed{}$

2 Write an equal number statement to match each array.

a $\boxed{} \times \boxed{} = \boxed{} \times \boxed{}$

b $\boxed{} \times \boxed{} = \boxed{} \times \boxed{}$

c $\boxed{} \times \boxed{} = \boxed{} \times \boxed{}$

d $\boxed{} \times \boxed{} = \boxed{} \times \boxed{}$

3 a Draw an array for a multiplication that uses any two different numbers from 3 to 7.

b Now write an equal number statement to match your array.

$\boxed{} \times \boxed{} = \boxed{} \times \boxed{}$

Date: _____

Lesson 4: **Solving problems (1)**

• Use arrays to solve 'real-life' multiplication problems

1 3 brothers have 5 toy cars each.

How many toy cars do they have altogether? ☐

2 There are 2 bowls, with 6 fish in each.

How many fish are there altogether? ☐

3 Solve the multiplication problems. Draw arrays in the boxes to help you.

   a Hassan picks plums for his 5 sisters to have 4 each.

   How many plums does he pick? ☐

   b Holly has 10 plant pots.
   She plants 2 seeds in each pot.

   How many seeds does she plant altogether? ☐

   c 7 learners have 2 crayons each.

   How many crayons are there altogether? ☐

4 David picks 10 flowers for each of his 3 aunts.

How many flowers does he pick altogether? ☐

5 Write an equal number statement that matches the problem in **4** .

☐ × ☐ = ☐ × ☐

Date: _____

Number

Lesson 1: **Division – sharing between 2**

• Share up to 20 objects between 2

1 Share the strawberries between 2 plates.

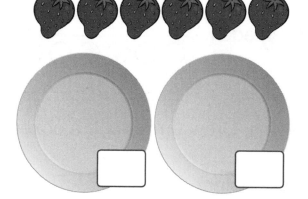

2 Share the sandwiches between 2 plates.

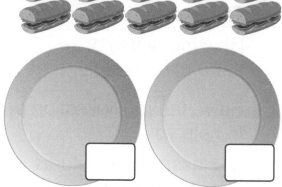

3 Share the cakes between 2 plates.

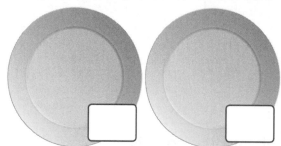

4 Share the tomatoes between 2 plates.

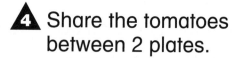

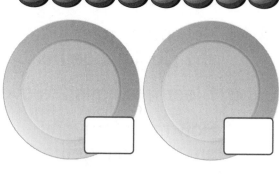

5 Share the cherries between 2 plates.

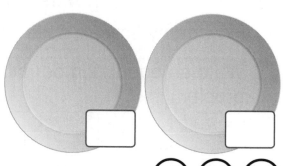

Date: _____

Number

Lesson 2: **Division – sharing between more than 2**

• Share amounts between more than 2

1 Share 6 toy cars equally between 3 children.

2 Share 20 bugs equally between 10 leaves.
How many bugs are on each leaf?

3 Share 30 sweets equally between 5 children.
How many sweets for each child?

4 Share 15 flowers equally between 3 vases.
How many flowers are in each vase?

5 Share 28 birds equally between 4 trees.
How many birds are in each tree?

Date: _____

Number

Lesson 3: **The division sign (1)**

• Recognise the division sign
• Share to solve division number sentences

1 Solve the divisions. Use the circles to help you.

a $6 \div 2 =$ ☐

b $15 \div 5 =$ ☐

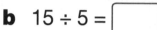

2 Solve the divisions. Use the sharing diagrams to help you.

a $16 \div 2 =$ ☐

X X X X X X X X X X X X X X X X

b $20 \div 10 =$ ☐

X X X X X X X X X X X X X X X X X X X X

c $35 \div 5 =$ ☐

XXXXXXXXXXXXXXXXXXXXXXXXXXXXXXXXX

d $24 \div 4 =$ ☐

X X X X X X X X X X X X X X X X X X X X X X X X

3 Solve the divisions. Draw sharing diagrams to help you.

a $40 \div 4 =$ ☐

b $27 \div 3 =$ ☐

Date: _____

Lesson 4: **Solving problems (2)**

- Solve 'real-life' division problems by sharing

You will need
- counters or a sharing diagram sheet

Use counters or a sharing diagram to solve the problems.

1 There are 8 ice creams and 4 children.

How many ice creams can each child have? ☐

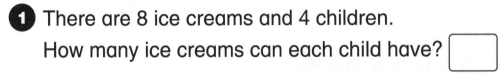

2 a There are 12 rabbits and 4 burrows. Share the rabbits equally. How many rabbits are in each burrow?

☐ ÷ ☐ = ☐

b There are 26 beads and 2 pieces of string. Share the beads equally. How many beads are on each necklace?

☐ ÷ ☐ = ☐

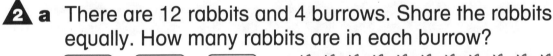

3 There are 65 biscuits in the tin and 5 plates. Share the biscuits equally. How many biscuits are on each plate?

☐ ÷ ☐ = ☐

Date: _____

Lesson 1: **Division as grouping (1)**

- Group objects to discover how many groups of 2 there are

Draw a ring around each group of 2 to find out how many groups there are.

1 a How many groups of 2 in 6? ☐

b How many groups of 2 in 8? ☐

2 a How many groups of 2 in 14? ☐

b How many groups of 2 in 10? ☐

c How many groups of 2 in 16? ☐

d How many groups of 2 in 12? ☐

3 Draw 18 socks. Draw rings around them in groups of 2.

How many groups of 2 in 18? ☐

Date: _____

Lesson 2: **Division as grouping (2)**

• Discover how many groups of 5 or 10 there are

1 Draw a ring around each group of 5 goats to find how many groups of 5 are in 10.

2 Draw groups of 10 dots to find how many groups of 10 are in 30.

3 Draw groups of 5 dots to find how many groups of 5 are in 25.

4 Draw groups of 5 dots to find how many groups of 5 are in 15.

5 Draw groups of 10 dots to find how many groups of 10 are in 60.

Date: _____

Number

Lesson 3: **The division sign (2)**

- Recognise the division sign
- Solve division number sentences by grouping

1 Solve the divisions. Draw rings around the groups to help you.

a $8 \div 2 = \boxed{}$

b $20 \div 5 = \boxed{}$

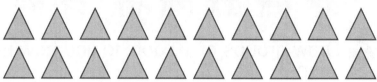

2 Solve the divisions. Draw groups to help you.

a $30 \div 5 = \boxed{}$

b $18 \div 2 = \boxed{}$

c $40 \div 10 = \boxed{}$

d $10 \div 2 = \boxed{}$

3 Solve the divisions. Draw groups to help you.

a $21 \div 3 = \boxed{}$

b $24 \div 4 = \boxed{}$

Date: _____

Lesson 4: **Solving problems (3)**

• Solve 'real-life' problems by grouping

You will need
• counters

Use counters or draw groups to solve the problems.

1 There are 8 children. The children get into pairs.

How many pairs are there? ☐

2 There are 20 sweets. Leo puts them into bags of 5.
How many bags of sweets are there?

☐ ÷ ☐ = ☐

3 Rosa has 12 fish. 4 fish live in each fish bowl. How many fish bowls does Rosa have?

☐ ÷ ☐ = ☐

4 A farmer has 21 goats. He can put 3 goats in each pen. How many pens does he need?

☐ ÷ ☐ = ☐

Date: _____

Lesson 1: **Division as repeated subtraction**

Number

> • Use objects to solve division problems with repeated subtraction

1 Subtract groups of 2 crabs by crossing them out until you have no crabs left. Then count how many groups of crabs you subtracted.

$10 \div 2 = \boxed{}$

2 Subtract groups of 5 fish by crossing them out until you have no fish left. Then count how many groups of fish you subtracted.

$35 \div 5 = \boxed{}$

3 Subtract groups of 3 seahorses by crossing them out until you have no seahorses left. Then count how many groups of seahorses you subtracted.

$18 \div 3 = \boxed{}$

4 Subtract groups of 4 starfish by crossing them out until you have no starfish left. Then count how many groups of starfish you subtracted.

$36 \div 4 = \boxed{}$

Date: _____

Lesson 2: **Division using a number line (1)**

> • Use a number line to solve division problems with repeated subtraction

1 Start on 6. Subtract groups of 2 until you reach 0. Count the jumps to find the answer.

$6 \div 2 =$ ⬚

2 Start on 8. Subtract groups of 2 until you reach 0. Count the jumps to find the answer.

$8 \div 2 =$ ⬚

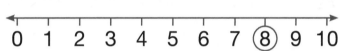

3 Subtract groups on the number line. Count the jumps to find the answer.

a $18 \div 2 =$ ⬚

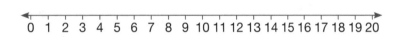

b $40 \div 5 =$ ⬚

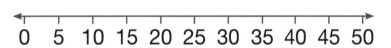

c $60 \div 10 =$ ⬚

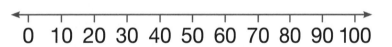

d $20 \div 2 =$ ⬚

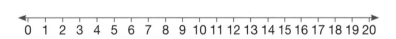

4 Subtract groups on the number line. Count the jumps to find the answer.

a $60 \div 5 =$ ⬚

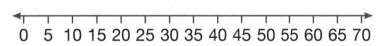

b $100 \div 10 =$ ⬚

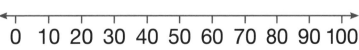

Date: _____

Number

Lesson 3: **Division using a number line (2)**

- Use a number line to solve division problems with repeated subtraction

1 Use the number lines to solve the divisions.

a 30 ÷ 10 = ☐

0   10   20   ⟨30⟩   40   50

b 10 ÷ 5 = ☐

0   5   ⟨10⟩   15   20

2 Use the number lines to solve the divisions.

a 70 ÷ 10 = ☐

0 10 20 30 40 50 60 70 80 90 100

b 25 ÷ 5 = ☐

0 5 10 15 20 25 30 35 40 45 50

c 16 ÷ 2 = ☐

0 1 2 3 4 5 6 7 8 9 10 11 12 13 14 15 16 17 18 19 20

d 50 ÷ 10 = ☐

0   10   20   30   40   50

3 Write a division number sentence to match the number lines.

a

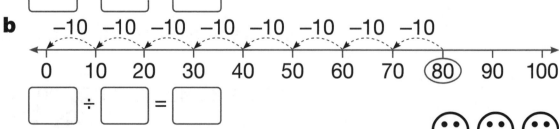

     –2   –2   –2   –2

0 1 2 3 4 5 6 7 ⟨8⟩ 9 10 11 12 13 14 15 16 17 18 19 20

☐ ÷ ☐ = ☐

b

  –10 –10 –10 –10 –10 –10 –10 –10

0   10   20   30   40   50   60   70   ⟨80⟩   90   100

☐ ÷ ☐ = ☐

Date: _____   ☺ ☺ ☹

Number

Lesson 4: **Solving problems (4)**

• Solve 'real-life' division problems using repeated subtraction

1 Use the number line to subtract groups to solve the problem. Lucy has 10 flowers. She gives each of her aunts 2 flowers.

How many aunts does Lucy have? ☐

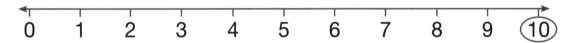

Draw number lines to subtract groups to solve the problems.

2 Miss Jones has a box of 30 pens. She puts 5 pens on each table. How many tables are there? ☐

3 James made 50 biscuits and put 10 biscuits in each box.

How many boxes does he have? ☐

4 Henry wants to give $2 each to some of his friends.

He has $30. How many friends can he give $2 to? ☐

Date: _____

Lesson 1: **2 times table: multiplication facts**

• Recall multiplication facts for the 2 times table

1 Write the multiplication facts to match the towers of 2 cubes.

a

b

$2 \times \boxed{} = \boxed{}$ $2 \times \boxed{} = \boxed{}$

2 Use the number line to answer the 2 times table facts.

1	2	3	4	5	6	7	8	9	10
2	4	6	8	10	12	14	16	18	20

a $2 \times 10 = \boxed{}$ **b** $2 \times 2 = \boxed{}$ **c** $2 \times 5 = \boxed{}$

d $2 \times 6 = \boxed{}$ **e** $2 \times 8 = \boxed{}$ **f** $2 \times 3 = \boxed{}$

3 Write the 2 times table fact to match the number on each star.

a ☆ 14 $\boxed{} \times \boxed{} = \boxed{}$

b ☆ 18 $\boxed{} \times \boxed{} = \boxed{}$

c ☆ 2 $\boxed{} \times \boxed{} = \boxed{}$

Date: _____

Lesson 2: **2 times table: division facts**

• Recall division facts for the 2 times table

1 Draw a ring around each group of 2 to complete the 2 times table division facts.

a

☐ ÷ 2 = ☐

b

☐ ÷ 2 = ☐

2 Use the number line to answer the 2 times table division facts.

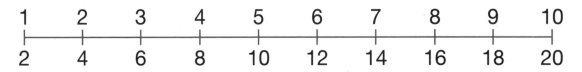

1	2	3	4	5	6	7	8	9	10
2	4	6	8	10	12	14	16	18	20

a 14 ÷ 2 = ☐   **b** 12 ÷ 2 = ☐   **c** 18 ÷ 2 = ☐

d 6 ÷ 2 = ☐   **e** 20 ÷ 2 = ☐   **f** 2 ÷ 2 = ☐

3 Write a 2 times table division fact and multiplication fact to match the number on each flag.

a 16 ☐ ÷ ☐ = ☐   ☐ × ☐ = ☐

b 4 ☐ ÷ ☐ = ☐   ☐ × ☐ = ☐

c 20 ☐ ÷ ☐ = ☐   ☐ × ☐ = ☐

Date: _____   ☺ ☺ ☹

51

Number

Lesson 3: **5 times table: multiplication facts**

- Recall multiplication facts for the 5 times table

1 Write the times table facts to match the towers of 5 cubes.

a

5 × ☐ = ☐

b
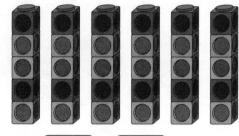

5 × ☐ = ☐

2 Use the number line to answer the 5 times table facts.

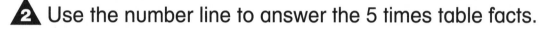

1	2	3	4	5	6	7	8	9	10
5	10	15	20	25	30	35	40	45	50

a 5 × 5 = ☐   **b** 5 × 10 = ☐   **c** 5 × 2 = ☐

d 5 × 8 = ☐   **e** 5 × 4 = ☐   **f** 5 × 9 = ☐

3 Write the 5 times table fact to match the number on each star.

a ☆ 45   ☐ × ☐ = ☐

b ☆ 20   ☐ × ☐ = ☐

c ☆ 35   ☐ × ☐ = ☐

Date: _____

Number

Lesson 4: **5 times table: division facts**

• Recall division facts for the 5 times table

1 Draw a ring around each group of 5 to complete the 5 times table division facts.

a

☐ ÷ 5 = ☐

b

☐ ÷ 5 = ☐

2 Use the number line to answer the 5 times table division facts.

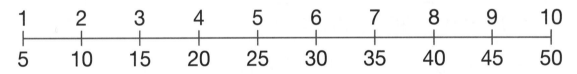

1	2	3	4	5	6	7	8	9	10
5	10	15	20	25	30	35	40	45	50

a 50 ÷ 5 = ☐   **b** 30 ÷ 5 = ☐   **c** 15 ÷ 5 = ☐

d 45 ÷ 5 = ☐   **e** 5 ÷ 5 = ☐   **f** 35 ÷ 5 = ☐

3 Write a 5 times table division fact and multiplication fact to match the number on each fish.

a  ☐ ÷ ☐ = ☐   ☐ × ☐ = ☐

b  ☐ ÷ ☐ = ☐   ☐ × ☐ = ☐

c  ☐ ÷ ☐ = ☐   ☐ × ☐ = ☐

Date: _____

53

Number

Lesson 1: **10 times table: multiplication facts**

• Recall multiplication facts for the 10 times table

1 Write the multiplication facts to match the tens.

a

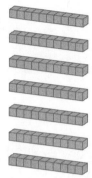

$10 \times \boxed{} = \boxed{}$

b

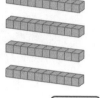

$10 \times \boxed{} = \boxed{}$

2 Use the number line to answer the 10 times table facts.

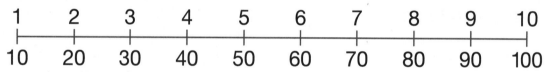

a $10 \times 10 = \boxed{}$ **b** $10 \times 5 = \boxed{}$ **c** $10 \times 1 = \boxed{}$

d $10 \times 3 = \boxed{}$ **e** $10 \times 8 = \boxed{}$ **f** $10 \times 9 = \boxed{}$

3 Write the 10 times table fact to match the number on each flag.

a 20 $\boxed{} \times \boxed{} = \boxed{}$

b 60 $\boxed{} \times \boxed{} = \boxed{}$

c 100 $\boxed{} \times \boxed{} = \boxed{}$

Date: _____

Lesson 2: **10 times table: division facts**

- Recall division facts for the 10 times table

1 Count the 10s to complete the 10 times table division facts.

a

☐ ÷ 10 = ☐

b

☐ ÷ 10 = ☐

c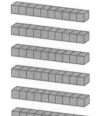

☐ ÷ 10 = ☐

2 Use the number line to answer the 10 times table division facts.

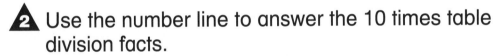
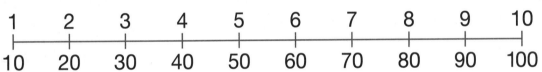

| 1 | 2 | 3 | 4 | 5 | 6 | 7 | 8 | 9 | 10 |

| 10 | 20 | 30 | 40 | 50 | 60 | 70 | 80 | 90 | 100 |

a 90 ÷ 10 = ☐   **b** 10 ÷ 10 = ☐   **c** 40 ÷ 10 = ☐

d 20 ÷ 10 = ☐   **e** 60 ÷ 10 = ☐   **f** 100 ÷ 10 = ☐

3 Write a 10 times table division fact and multiplication fact to match the number on each fish.

a ☐ ÷ ☐ = ☐   ☐ × ☐ = ☐

b ☐ ÷ ☐ = ☐   ☐ × ☐ = ☐

c 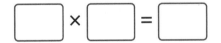 ☐ ÷ ☐ = ☐   ☐ × ☐ = ☐

Date: _____   ☺ 😐 ☹

Number

Lesson 3: **1 times table: multiplication and division facts**

- Recall multiplication and division facts for the 1 times table

1 Write the multiplication and division facts to match the ones.

a

$1 \times 3 =$ ☐

b

$5 \div 1 =$ ☐

2 Use the number line to answer the 1 times table multiplication and division facts.

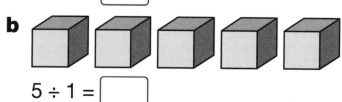

a $1 \times 4 =$ ☐   **b** $1 \times 10 =$ ☐   **c** $1 \times 8 =$ ☐

d $1 \div 1 =$ ☐   **e** $9 \div 1 =$ ☐   **f** $2 \div 1 =$ ☐

3 Write 1 times table multiplication and division number sentences to match the number on each star.

a ☆6 ☐ × ☐ = ☐   ☐ ÷ ☐ = ☐

b ☆9 ☐ × ☐ = ☐   ☐ ÷ ☐ = ☐

c ☆7 ☐ × ☐ = ☐   ☐ ÷ ☐ = ☐

Date: _____

Lesson 4: **1, 2, 5 and 10 times tables**

- Recall multiplication and division facts for the 1, 2, 5 and 10 times tables

You will need
- coloured pencils

1 Colour each answer on the times table grid.

a 5×3

b 10×9

c $16 \div 2$

d $70 \div 10$

e 2×6

f $20 \div 5$

×	1	2	3	4	5	6	7	8	9	10
1	1	2	3	4	5	6	7	8	9	10
2	2	4	6	8	10	12	14	16	18	20
5	5	10	15	20	25	30	35	40	45	50
10	10	20	30	40	50	60	70	80	90	100

2 Use the times table grid in **1** to answer the facts.

a $10 \times 2 =$ ☐　　　**b** $2 \times 4 =$ ☐　　　**c** $25 \div 5 =$ ☐

d $14 \div 2 =$ ☐　　　**e** $5 \times 9 =$ ☐　　　**f** $80 \div 10 =$ ☐

g $1 \times 5 =$ ☐　　　**h** $50 \div 10 =$ ☐　　　**i** $35 \div 5 =$ ☐

j $10 \times 4 =$ ☐　　　**k** $10 \times 5 =$ ☐　　　**l** $9 \div 1 =$ ☐

3 Write a number sentence for each problem.

a 10 children have 3 balls each. How many balls is this?

b A teacher shares 50 crayons between 5 tables. How many crayons does each table get?

c 2 cats catch 5 mice each. How many mice do they catch?

Date: _____　　😊 😐 ☹

Lesson 1: **Recognising local currency**

- Recognise the currency symbol in local currency
- Recognise the value of coins and notes in local currency

You will need
- selection of coins or notes from your local currency
- coloured pencils

1 Colour the symbols used in your local currency. If you cannot see your local currency symbol, draw it here.

€ $ £ ¥ ₹ ₽ ฿ ₴ ₩ đ ₸ ₵

2 Draw two coins from your local currency. Make sure that you include the currency (if it's shown) and the value of each coin.

3 Draw a note from your local currency. Make sure that you include the currency (if it's shown) and the value of the note.

4 Draw the coin or note with the lowest value in your currency. Draw the coin or note with the highest value in your currency. Make sure that you include the currency (if it's shown) and the value of the coin or note.

lowest

highest

Date: _____

Number

Lesson 2: **Paying with dollars and cents**

- Recognise all dollar and cent coins and notes
- Match values of coins and notes to prices

1 Fill in the value of each coin or note.

a ☐ c **b** ☐ c **c** $ ☐

2 Draw a line to match each coin and note to the item with the same value.

50c

1c

$10

$1

10c

3 You have a $5 note. Draw a ring around any toys that cost **too much** for you to buy.

50c $2 $5 $25 $10

Date: _____

Number

Lesson 3: **Comparing values**

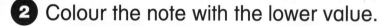

• Compare and order values of coins and notes

1 Colour the coin with the higher value.

2 Colour the note with the lower value.

3 In each set, colour the coin or note with the highest value. Draw a ring around the coin or note with the lowest value.

a

b

c

d

4 Order the coins and notes from lowest value (1) to highest value (5).

Date: _____

Lesson 4: **Equal values**

• Find coins or notes of the same value

1 Is the value of the coins and notes the same or different?
Draw a ring around your answer.

a

b

same   different

same   different

c

same   different

2 Draw a line to match the total in each purse with the coin or note.

3 Show another way to make $1.

Date: _____

☺ 😐 ☹

61

Number

Lesson 1: **Tens and ones**

• Know how many tens and ones are in a 2-digit number

1 How many tens? How many ones?

a

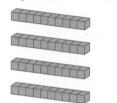

tens: [ ]   ones: [ ]

b

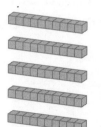

tens: [ ]   ones: [ ]

2 How many tens? How many ones?

a 57

tens: ones:
[ ] [ ]

b 30

tens: ones:
[ ] [ ]

c 61

tens: ones:
[ ] [ ]

d 95

tens: ones:
[ ] [ ]

e 44

tens: ones:
[ ] [ ]

f 28

tens: ones:
[ ] [ ]

3 Write a 2-digit number. [ ]

How many tens? [ ]

How many ones? [ ]

Draw Base 10 for your number.

Date: _____

Lesson 2: **Partitioning**

• Partition 2-digit numbers into tens and ones

1 Complete the partitioning.

a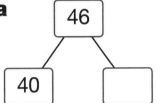

b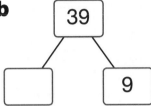

c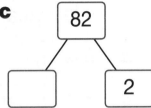

2 Partition these numbers.

a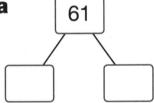

b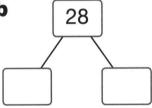

c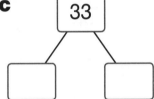

d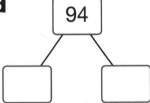

e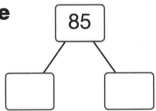

f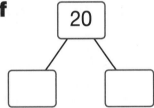

3 Fill in the missing numbers.

a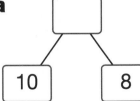

b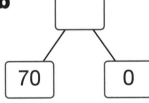

c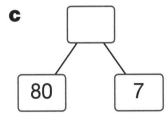

Date: _____

Number

Lesson 3: **Comparing numbers**

• Compare 2-digit numbers

You will need
• red and blue coloured pencils

1 Colour the correct word.

a 24 is [greater] [less] than 20.

b 78 is [greater] [less] than 79

c 53 is [greater] [less] than 55.

2 Colour the greatest number.

a (62) (23) (84)

b (96) (42) (70)

c (83) (93) (73) (89)

3 Colour the smallest number.

a (18) (28) (68)

b (61) (75) (53)

c (45) (30) (50) (40)

4 Colour the greatest number red and the smallest number blue.

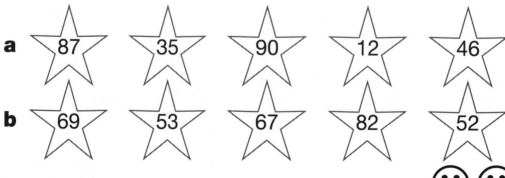

a 87   35   90   12   46

b 69   53   67   82   52

Date: _____

Lesson 4: **Ordinal numbers**

• Use ordinal numbers

You will need
• blue, red and yellow coloured pencils

1 Fill in the missing ordinal numbers.

1st   2nd   ☐   4th   ☐   ☐   7th   8th   ☐   10th

2

1st

a Colour the 7th flower blue.

b Colour the 10th flower red.

c Colour the 3rd flower yellow.

3 Fill in the missing dates.

4 The day **after** the 11th is party day. Colour it red.

May						
M	T	W	T	F	S	S
					1st	
3rd	4th		6th			9th
10th		12th	13th	14th	15th	16th
17th	18th		20th		22nd	23rd
	25th	26th		28th		30th

Date: _____

Lesson 1: **Composing and decomposing numbers**

Number

• Compose and decompose numbers using tens and ones

1 Complete the number sentences.

a $50 + \boxed{} = 56$ **b** $\boxed{} + 3 = 83$ **c** $20 + \boxed{} = 29$

2 Write a number sentence to match each number.

a $\boxed{} + \boxed{} = 75$ **b** $\boxed{} + \boxed{} = 91$

c $\boxed{} + \boxed{} = 40$ **d** $39 = \boxed{} + \boxed{}$

e $64 = \boxed{} + \boxed{}$ **f** $22 = \boxed{} + \boxed{}$

3 Write two number sentences for each number.

a 68

$\boxed{} + \boxed{} = \boxed{}$
$\boxed{} = \boxed{} + \boxed{}$

b 27

$\boxed{} + \boxed{} = \boxed{}$
$\boxed{} = \boxed{} + \boxed{}$

c 50

$\boxed{} + \boxed{} = \boxed{}$
$\boxed{} = \boxed{} + \boxed{}$

d 94

$\boxed{} + \boxed{} = \boxed{}$
$\boxed{} = \boxed{} + \boxed{}$

Date: _____

Number

Lesson 2: **Regrouping numbers**

• Regroup 2-digit numbers

1 Complete each number sentence.

   a $20 + \boxed{} = 26$

   b $20 + \boxed{} + \boxed{} = 26$

   c $20 + \boxed{} + \boxed{} + \boxed{} = 26$

2 Complete each number sentence.

   $64 = 60 + \boxed{}$ $64 = 60 + \boxed{} + \boxed{}$

3 Write a different way to show 64.

4 Complete each number sentence.

   $43 = 40 + \boxed{}$ $43 = \boxed{} + \boxed{} + \boxed{} + \boxed{} + 3$

5 Write a different way to show 43.

6 Regroup 35 in as many ways as you can. Try regrouping the ones, then the tens.

Date: _____

Lesson 3: **Ordering 2-digit numbers**

Number

• Compare and order 2-digit numbers

1 Order each set of numbers, starting with the **smallest**.

a 72   81   24

☐ ☐ ☐

b 16   36   26

☐ ☐ ☐

2 Order each set of numbers, starting with the **smallest**.

a 83   99   27   40   62   **b** 53   91   30   29   68

☐ ☐ ☐ ☐ ☐   ☐ ☐ ☐ ☐ ☐

3 Order each set of numbers, starting with the **greatest**.

a 33   52   81   27   75   **b** 84   48   56   90   53

☐ ☐ ☐ ☐ ☐   ☐ ☐ ☐ ☐ ☐

4 Write each number on the number line.

a

☆ 78

50   60   70   80   90   100

b

☆ 33

0   10   20   30   40   50

5 Order the numbers on the number line.   ☆ 85   ☆ 61   ☆ 52   ☆ 97   ☆ 79   ☆ 91

50   60   70   80   90   100

Date: _____

Number

Lesson 4: **Rounding 2-digit numbers**

- Round 2-digit numbers to the nearest 10

1 Draw lines to show which 10 each number rounds to.

24   29   25

20   28   23   27   **30**

22   21   26

2 Round these numbers to the nearest 10.

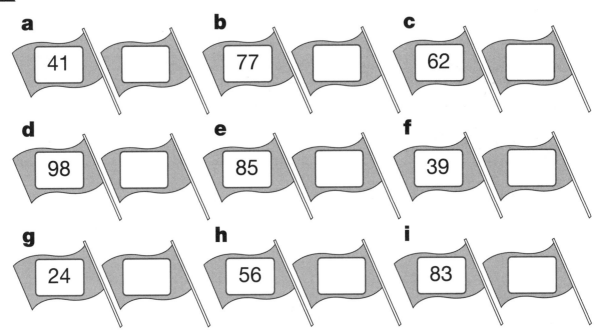

a 41 ☐   **b** 77 ☐   **c** 62 ☐

d 98 ☐   **e** 85 ☐   **f** 39 ☐

g 24 ☐   **h** 56 ☐   **i** 83 ☐

3 Write six numbers that round to 70.

☐ ☐ ☐ 70 ☐ ☐ ☐

4 Write eight numbers that round to 50.

☐ ☐ ☐ ☐ 50 ☐ ☐ ☐ ☐

Date: _____

Number

Lesson 1: **Quarters of shapes**

• Recognise which shapes are divided into quarters and which shapes are not

You will need
• coloured pencil

1 Colour the shapes that are divided into quarters.

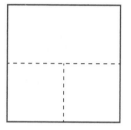
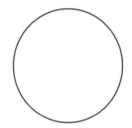
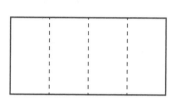

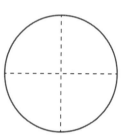
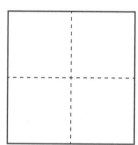

2 Colour the shapes that are divided into quarters. Then label each quarter $\frac{1}{4}$.

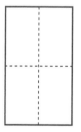
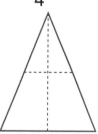
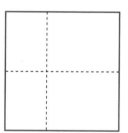
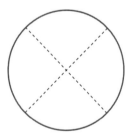

3 Colour the shapes that have been labelled correctly.

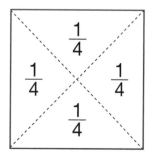
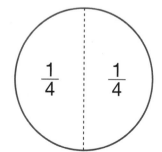

Date: _____

Lesson 2: **Finding one quarter**

• Find one quarter of a shape

You will need
• coloured pencil

1 Colour $\frac{1}{4}$ quarter of each shape.

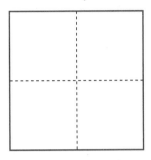 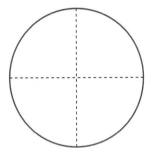 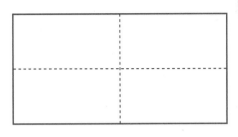

2 Draw lines to divide these shapes into quarters. Then colour $\frac{1}{4}$ of each shape.

  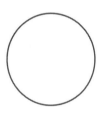 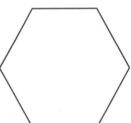

3 $\frac{1}{4}$ of each shape is shown. Draw the rest of the shape.

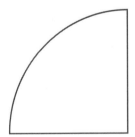 

Date: _____

Number

Number

Lesson 3: **Quarters of sets of objects**

• Find one quarter of a set of objects

1 How many stars are in each quarter?

a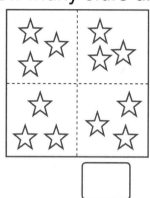

b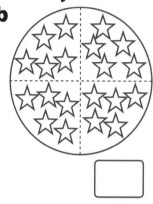

c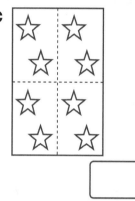

2 Draw dots in each quarter to find $\frac{1}{4}$.

a There are 12 children. $\frac{1}{4}$ of them go to the cinema.

How many children is that? ☐

b Asim has 8 pets. $\frac{1}{4}$ of his pets are rabbits.

How many rabbits does Asim have? ☐

c Priya has 20 strawberries.
Her mum says she may eat $\frac{1}{4}$ of them.

How many strawberries can Priya eat? ☐

3 Write a number story to match this number sentence. Then work out the answer. $\frac{1}{4}$ of 16 = ☐

_____

_____

Date: _____

Lesson 4: **Finding one quarter of a set of objects**

- Use one quarter of a set of objects to work out how many objects are in the whole set

1 Draw the rest of the counters. Then count how many there are altogether to find how many are in the full set.

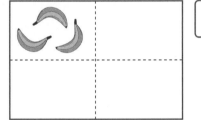	draw 1 counter
draw 1 counter	draw 1 counter

2 Draw the rest of the fruit. Then count how many there are altogether to find how many are in the full set.

a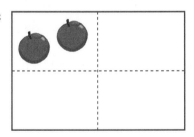

b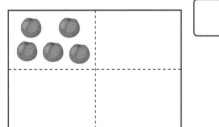

c

d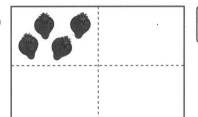

3 There are 7 marbles in one quarter of Ben's marble collection. How many marbles does he have altogether?

Draw the marbles to help you.

Date: _____

Number

Lesson 1: **Equal fractions**

- Recognise how big fractions are, compared to each other
- Recognise fractions that equal the same amount

1 Colour $\frac{1}{2}$ of each shape.

You will need
- coloured pencil

 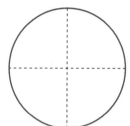 

2 Draw lines to join the shapes that have an equal fraction shaded.

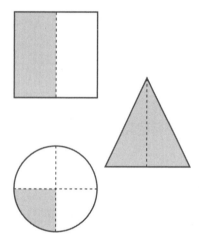 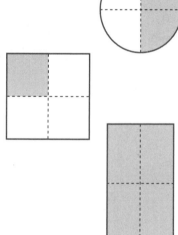

3 Draw a ring around the fractions that match the square.

$\frac{1}{2}$ 1 $\frac{2}{2}$ $\frac{2}{4}$ $\frac{1}{4}$ $\frac{4}{4}$

4 A cake is cut into four quarters. Lily gets half the cake and Henry gets two quarters of the cake. Who has more cake? Explain your answer.

_____

_____

Date: _____

Number

Lesson 2: **Combining fractions**

• Combine halves and quarters to create new fractions

You will need
• coloured pencil

1 Draw lines to match each diagram to a fraction.

| $\frac{1}{2}$ | |

| $\frac{1}{4}$ | | | |

| $\frac{1}{2}$ | $\frac{1}{2}$ |

$\frac{1}{4}$　　　　　　　1　　　　　　　$\frac{1}{2}$

2 Draw lines to match each diagram to a fraction.

| $\frac{1}{2}$ | $\frac{1}{4}$ | |　•

| $\frac{1}{2}$ | $\frac{1}{2}$ |
| $\frac{1}{2}$ | |　•

| $\frac{1}{4}$ | $\frac{1}{4}$ | $\frac{1}{4}$ | |　•

| $\frac{1}{4}$ | $\frac{1}{4}$ | | |　•

• 1 and $\frac{1}{2}$

• $\frac{3}{4}$

• $\frac{1}{2}$

• $\frac{3}{4}$

3 Colour the diagrams to show the fractions.

a

| | |
| | | |

1 and $\frac{1}{4}$

b

| | | | |
| | | | |

1 and $\frac{3}{4}$

Date: _____

75

Number

Lesson 3: **Fractions as operators**

- Find one half or one quarter of a shape or an amount to 20

You will need
- coloured pencil

1 Colour $\frac{1}{2}$ of the rectangle.

2 Colour $\frac{1}{4}$ of the circle.

3 Colour $\frac{1}{2}$ of the stars.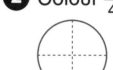

4 Colour $\frac{1}{2}$ of the triangles.

5 Colour $\frac{1}{2}$ of the hexagons.

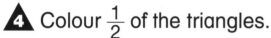

6 Colour $\frac{1}{4}$ of the squares.

7 Colour $\frac{1}{4}$ of the stars.

8 Colour $\frac{1}{4}$ of the triangles.

9 $\frac{1}{4}$ of 16 = ☐. Draw dots to help you.

10 $\frac{1}{4}$ of 24 = ☐. Draw dots to help you.

Date: _____

Lesson 4: **Fractions as division**

• Understand fractions as division

1 Use the stars to complete the number sentences.

a ☆☆☆☆☆☆

$\frac{1}{2}$ of 6 = ☐

☐ ÷ ☐ = ☐

b ☆☆☆☆☆☆☆☆

$\frac{1}{4}$ of 8 = ☐

☐ ÷ ☐ = ☐

2 Complete the number sentences to solve the problems.

a There are 10 cherries in a bowl. Jo eats half of them. How many cherries does Jo eat?

☐ ÷ ☐ = ☐

b 12 kittens are in a basket. Half of them are asleep. How many kittens are asleep?

☐ ÷ ☐ = ☐

c There are 8 biscuits in a packet. Tom eats one quarter of them. How many biscuits does Tom eat?

☐ ÷ ☐ = ☐

d There are 12 flowers in a vase. One quarter of them are red. How many red flowers are in the vase?

☐ ÷ ☐ = ☐

3 Your friend has 16 marbles. You win 8 of them in a game. What fraction of the marbles do you have? ☐

Write the fraction statement. ☐ of ☐ = ☐

4 20 birds were sitting on a roof. 5 of them flew away. What fraction of the birds flew away? ☐

Write the fraction statement. ☐ of ☐ = ☐

Date: _____

(side tab) **Geometry and Measure**

Lesson 1: **Using a calendar**

• Use and interpret a calendar

1 Fill in the missing days of the week and dates in January.

January						
Monday			**Thursday**		**Saturday**	
		1	2 piano lesson	3	4	5
	7	8	9		11 swimming	football
13	14	15 beach party	16 piano lesson	17	18 tennis lesson	19 football
	21		23 piano lesson	24 swimming	25 barbecue	26 football
27	swimming	29	piano lesson	31		

2 Use the calendar in **1** to answer these questions.

a How many days in January? _____

b What day of the week is 3rd January? _____

c What is happening on 15th January? _____

d How many times is swimming on the calendar? _____

e What day of the week are piano lessons? _____

3 Use the calendar in **1** to answer these questions.

a If you wanted to go for a 7 day holiday,
on which dates are you free? [   ] until [   ]

b If a friend wants to come and stay on a
Saturday, on which date could they come? [   ]

c Could you join the cricket club that is on Monday
afternoons?

Yes   **No**   Why? _____

Date: _____   ☺ ☺ ☹

Lesson 2: **Ordering time**

• Order units of time

You will need

• blue and red coloured pencils

1 Underline the month it is now in red. Underline the first month of the year in blue. Draw a ring around the last month of the year.

January   February   March   April   May   June   July
August   September   October   November   December

2 Which month comes next?

| January | February | March | |

| September | October | November | |

3 Number these units of time from shortest (1) to longest (7).

| day | | hour | | minute | |

| year | | month | | week | |

| second | |

4 Write something that takes about 1 second.

_____

5 Write something that takes about 1 minute.

_____

6 Write something that takes about 1 hour.

_____

Date: _____

Geometry and Measure

Geometry and Measure

Lesson 3: **Reading and showing the time (analogue)**

- Read and record the time to 5 minutes on an analogue clock

1 Count in fives around the clock to show how many minutes past and to each hour it would be.

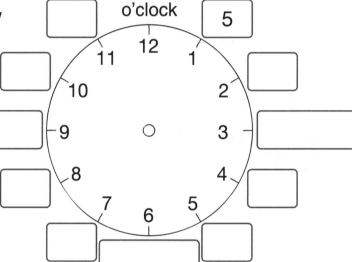

TO   o'clock   PAST   5

2 Write the time.

a

_____

b

_____

c

_____

d

_____

3 Draw the times on the clocks.

a 20 past 7

b 25 to 12

Date: _____

Lesson 4: **Reading and showing the time (digital)**

Geometry and Measure

• Read and record the time to 5 minutes on a digital clock

1 Draw a ring around the digital clocks.

2 Write the digital times to match.

a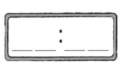

b

3 Write the digital times to match.

a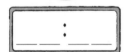

b

c

d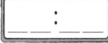

4 Write the digital time that will be 5 minutes **after** each time.

a

b

Date: _____

Lesson 1: **2D shapes (1)**

Geometry and Measure

• Identify, describe and sort 2D shapes

1 Write the letter of each shape in the correct pot.

A   B   C
D   E   F   G

hexagon   **not a hexagon**

2 Colour the pentagons and hexagons to find your way home.

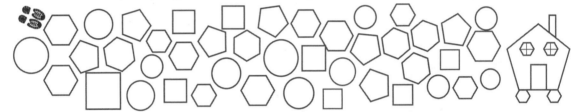

3 Sort the shapes by writing the letters in the correct box.

A   B   C

D   E

F   G

4 or more vertices	3 or fewer vertices

4 Sort the shapes by writing the letters in the correct boxes.

A   B   C   D   E   F   G

4 or fewer sides	5 or more sides	4 or fewer vertices

Date: _____

Geometry and Measure

Lesson 2: **2D shapes (2)**

- Identify 2D shapes in familiar objects
- Explore patterns using shapes

You will need
- geoboard
- elastic bands

1 Use the geoboard to make a **square** and a **triangle**. Then copy them below.

2 Use the geoboard to make three different shapes that you can see around the classroom. Then copy them below.

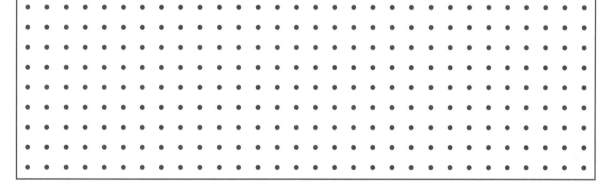

3 Use the geoboard to make a **pentagon** and a **hexagon**. Then draw them in a repeating pattern below.

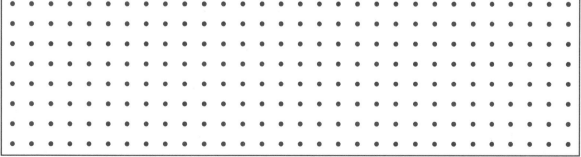

Date: _____

Lesson 3: **Lines of symmetry**

Geometry and Measure

• Recognise horizontal and vertical lines of symmetry

You will need
• mirror

1 Colour the shapes that have symmetry. Use a mirror to help.

2 Colour the shapes that have both vertical and horizontal lines of symmetry. Use a mirror to help.

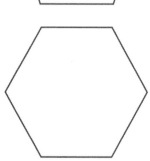

h o r i z o n t a l

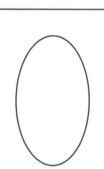

v e r t i c a l

3 Use a mirror to find how many lines of symmetry there are in each shape. Write the number in the box.

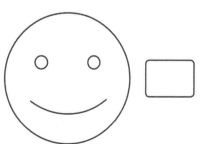 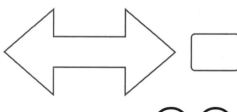

Date: _____

84

Lesson 4: **Angles**

- Identify right angles
- Understand that a right angle is a quarter turn

1 Draw a ring around the right angle.

2 Draw a ring around the angle that is **not** a right angle.

3 Make a list of things you can see in the classroom that have right angles.

- _____
- _____
- _____
- _____
- _____

4 Draw a ring around the person who has made a right angle with their turn.

half turn    full turn    quarter turn

Date: _____

Lesson 1: **Recognising 3D shapes**

Geometry and Measure

- Identify and name 3D shapes
- Recognise 3D shapes in familiar objects

You will need
- red and blue coloured pencils

1 Shade the 2D shapes red and the 3D shapes blue.

2 Match each object to its shape.

   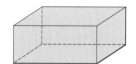

3 Draw a ring around the correct 3D shape name for each row.

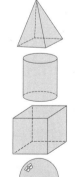

triangle	hexagon	pyramid	cube
sphere	cuboid	cylinder	circle
pentagon	cube	rectangle	square
cylinder	sphere	triangle	cuboid
hexagon	cube	circle	cuboid

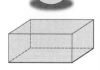

Date: _____

Lesson 2: **Describing 3D shapes**

- Talk about the faces, edges and vertices of 3D shapes

You will need
- red and blue coloured pencils

1 Trace over each edge in red. Draw a blue dot on each vertex.

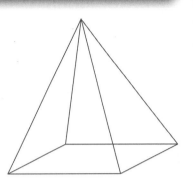

2 Complete the table.

A   B 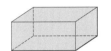   C   D

	Name of shape	Number of faces	Number of vertices	Number of edges
A				
B				
C				
D				

3 Draw a ring around the 3D shape.

Which shape am I? I have 3 faces, 2 edges and no vertices.

Date: _____

Geometry and Measure

Lesson 3: **Sorting 3D shapes**

Geometry and Measure

• Compare and sort 3D shapes

You will need
• red and blue coloured pencils

1 Colour red all the shapes with 3 or fewer faces.

Colour blue all the shapes with 4 or more faces.

    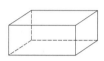

2 Colour red all the shapes with fewer than 8 edges.
Colour blue all the shapes with 8 or more edges.

  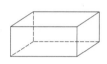

3 Colour red all the shapes with fewer than 5 vertices.
Colour blue all the shapes with 5 or more vertices.

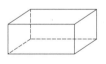

4 Colour the shapes that fit each sorting rule.

Curved faces only

Flat faces only

Curved and flat faces

Date: _____

Lesson 4: **Making 3D shapes**

• Make 3D shapes

You will need

• paper, scissor, glue stick or sticky tape, straws and modelling clay

1 Use paper to make a cylinder.
Draw your cylinder.

2 Use straws and modelling clay to make a pyramid.
Draw your pyramid.

3 Use straws and modelling clay to make a cuboid.
Draw your cuboid.

Date: _____

Lesson 1: **Measuring length with non-standard units**

Geometry and Measure

- Estimate and measure length with units of measure that are the same

You will need
- small counters
- interlocking cubes

1 How many cubes does each dinosaur bone measure?

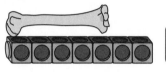

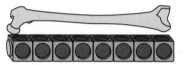

2 Estimate the length of each pencil using small counters. Then use the counters to measure each pencil.

a Estimate: ☐ Length: ☐

b Estimate: ☐ Length: ☐

c Estimate: ☐ Length: ☐

d Estimate: ☐ Length: ☐

3 Estimate the length in cubes of pencil **a** and pencil **b** in **2**. Then use cubes to measure each pencil

a Estimate: ☐ Length: ☐

b Estimate: ☐ Length: ☐

4 Which is better to measure the length of a pencil: paper clips or cubes? _____

Why? _____

Date: _____

Lesson 2: **Centimetres and metres**

• Recognise and use the standard units: centimetres and metres

You will need
• coloured pencils

Geometry and Measure

1 Tick the things that are **longer** than a 30 centimetre ruler. Draw a ring around the things that are **shorter** than a 30 centimetre ruler.

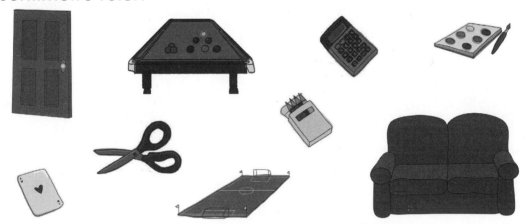

2 Draw one example in each box.

Shorter than 30 centimetres	Longer than 30 centimetres	Longer than 1 metre

3 Shade a 1 centimetre length on the ruler and write in the missing numbers.

12

Date: _____

Lesson 3: **Drawing and measuring lines**

Geometry and Measure

• Draw and measure lines with a ruler

You will need
• ruler

1 Use your ruler to measure these lines.

a ——————

b ——————————————

2 Draw a line 4 centimetres long.

3 Draw a line 10 centimetres long.

4 Draw a line 13 centimetres long.

5 Draw a line 6 centimetres long.

6 Draw a line that is 3 centimetres **longer** than this line.

——————————————

7 Draw a line that is 2 centimetres **shorter** than this line.

——————————————

Date: _____

Lesson 4: **Estimating length**

- Estimate lengths of objects in centimetres

You will need
- ruler

1 Estimate the length of the centipede, then measure it.

Estimate: ☐   Length: ☐

2 Estimate the length of the beetle, then measure it.

Estimate: ☐   Length: ☐

3 Estimate the height of each candy stick, then measure it.

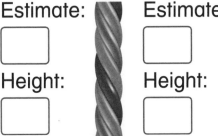

Estimate: ☐   Estimate: ☐   Estimate: ☐   Estimate: ☐

Height: ☐   Height: ☐   Height: ☐   Height: ☐

4 Draw a pencil that you estimate to be 12 centimetres long. Then measure your drawing of the pencil to find out how long it really is.

Date: _____

Geometry and Measure

Geometry and Measure

Lesson 1: **Measuring mass with non-standard units**

You will need
- balance scale
- two objects to find the mass of
- pile of small items, such as cubes or marbles
- coloured pencils

- Estimate and measure mass with units of measure that are the same

1 Draw a ring around the lighter object on each balance scale.

2 Find the mass of two different objects. Use the balance scales to show what you did. Estimate the mass of each object first.

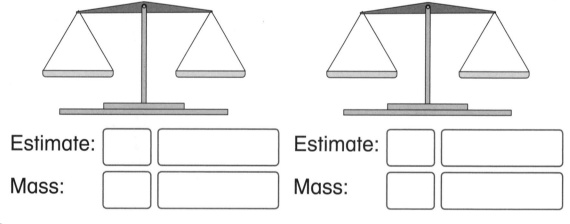

Estimate: ☐ ☐   Estimate: ☐ ☐

Mass: ☐ ☐   Mass: ☐ ☐

3 Draw the crayons needed to make each scale balance.

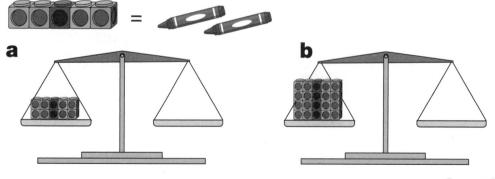

a   **b**

Date: _____

Lesson 2: **Grams and kilograms**

• Recognise and use the standard units: grams and kilograms

1 Draw a ring around the objects **heavier** than a kilogram.

2 Draw a ring around the measurement you would use for each object.

grams · kilograms　　grams · kilograms　　grams · kilograms　　grams · kilograms

grams · kilograms　　grams · kilograms　　grams · kilograms　　grams · kilograms

3 What is the mass of each item of food?

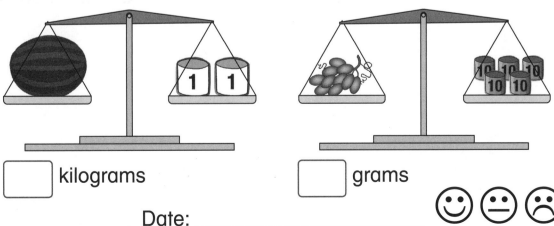

☐ kilograms　　☐ grams

Date: _____　　☺ 😐 ☹

95

Lesson 3: **Measuring mass with grams and kilograms**

Geometry and Measure

• Measure mass in grams and kilograms

1 What is the mass of each fruit?

a

[ ] grams

b

[ ] grams

c

[ ] kilograms

2 What is the mass of each bag of food?

a

[ ] kilograms

b

[ ] kilograms

c

[ ] grams

3 Show the mass on the scales.

a 40 grams

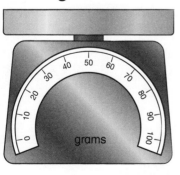

b 35 grams

c 9 kilograms

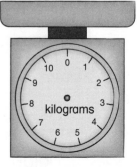

Date: _____

Lesson 4: **Comparing mass**

- Compare mass in grams and kilograms

1 Write the mass of each plate of cakes. Then complete
the sentences.

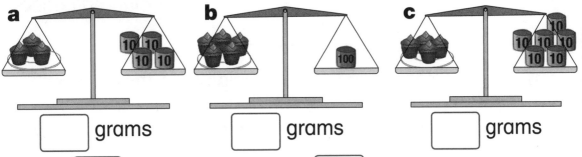

a | | grams   b | | grams   c | | grams

Plate | | is the lightest. Plate | | is the heaviest.

2 Write the mass of each box of food. Then complete
the statement.

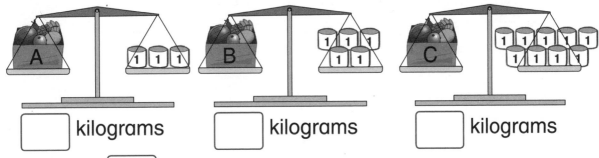

| | kilograms | | kilograms | | kilograms

Box C is | | kilograms heavier than Box A.

3 Write the mass of each box of bottles. Then complete
the sentence.

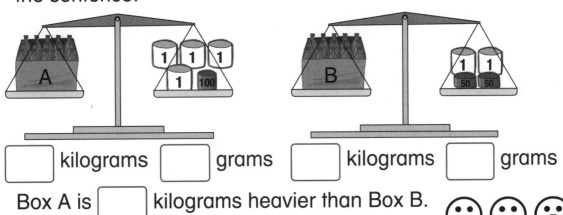

| | kilograms | | grams | | kilograms | | grams

Box A is | | kilograms heavier than Box B.

Date: _____

Geometry and Measure

Lesson 1: **Measuring capacity with non-standard units**

• Estimate and measure capacity with units of measure that are the same

You will need
• jug and small cup
• water
• coloured pencils

1 Draw a ring around the correct phrase.

a   holds **more than**   holds **less than**

b   holds **more than**   holds **less than**

c   holds **more than**   holds **less than**

2 Take a jug and a cup. Draw them below. Estimate how many cups of water the jug will hold. Then find out.

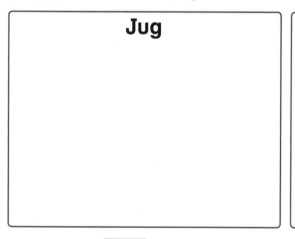

Jug	Cup

Estimate: ☐ cups   Capacity: ☐ cups

3 What is the capacity of one small bottle in tablespoons?

 holds the same as   holds the same as ☐ tablespoons

Date: _____

Lesson 2: **Litres and millilitres**

- Recognise and use the standard units: litres and millilitres

You will need
- coloured pencil

1 Draw a ring around the containers that hold **less than** 1 litre.

Tick the containers that hold **more than** 1 litre.

2 Colour each container to show how many litres are in it.

a 3 litres

b 5 litres

c 1 litre

litre   litre   litre

3 Colour each container to show how many millilitres are in it.

a 70 millilitres

b 20 millilitres

c 40 millilitres

millilitre   millilitre   millilitre

4 How many litres of water altogether? ☐ litres

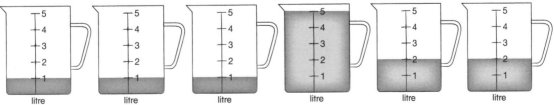

litre   litre   litre   litre   litre   litre

Date: _____

Lesson 3: **Measuring capacity using litres and millilitres**

Geometry and Measure

- Estimate and measure capacity in litres and millilitres

You will need
- containers of different sizes
- 1 litre measuring jug
- water
- coloured pencil

1 Look at your containers. Draw them below to show how much water you think each one holds.

I think these containers will hold **less than** 1 litre.	I think these containers will hold **more than** 1 litre.

2 Look at your containers. Draw them below to show how much water each one holds.

Holds **less than** half a litre	Holds about half a litre	Holds **more than** half a litre

3 Look at the containers that hold **less than** 1 litre.
How many millilitres does each of these containers hold?

Date: _____

Lesson 4: **Measuring temperature**

- Compare temperatures
- Read and interpret the scale on a thermometer

1 Draw a ring around the highest temperature. Tick the lowest temperature.

2 Write each temperature to the nearest 10 degrees.

a

b

c

d

degrees   degrees   degrees   degrees

3 Describe the temperature on each thermometer.

a

_____

_____

_____

_____

b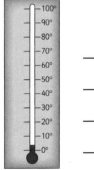

_____

_____

_____

_____

Date: _____

Geometry and Measure

Lesson 1: **Drawing reflections**

• Sketch the reflection of a 2D shape

You will need
• mirror

1 Sketch the reflection of
the square.
Use a mirror to help you.

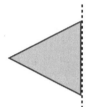

 Sketch the reflection of each shape. Use a mirror to help you.

a    **b**   **c**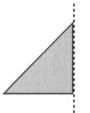

3 Predict what the pentagon will look
like reflected. Sketch your prediction.

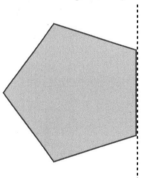

4 Now use a mirror to check.
Sketch the reflection.

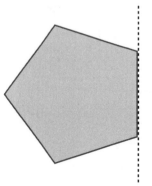

Date: _____

Lesson 2: **Position**

• Give and follow instructions for position

1 Draw a ring around the correct position of Ted.

a    **b**    **c**

next to   behind   on top

between   in front   under

2 Use the words in the box to complete the sentences.

above	next to	in front of

The cows are _____ the trees.

The calf stands _____ its mum.

The sun is _____ the trees.

3 Draw each creature to show its position in the picture.

a **On top of** the mushroom is a

b Hanging **next to** the jar is a

c **In front of** the jar is a

Date: _____

Geometry and Measure

Lesson 3: **Direction and movement**

Geometry and Measure

> • Follow and give directions to move from one position to another

1 Draw lines to match the directions to the arrows.

forward   left   right

2 Draw two rocks to block the way between the rabbit and its burrow. Write directions to get the rabbit home.

Directions:

_____

_____

_____

_____

3 Help the cat get to the milk by giving it directions. Use the code below.
The first two have been done for you.

Forward = F   Right = R   Left = L

F 3 _____   _____

Turn R, then F 2 _____   _____

_____   _____

_____   _____

_____   _____

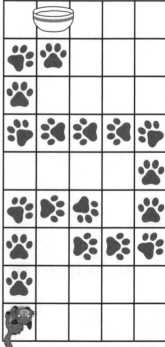

Date: _____

Lesson 4: **Whole, half and quarter turns**

- Make whole, half and quarter turns clockwise and anticlockwise

You will need
- coloured pencils

1 Draw lines to match the directions to the circles.

clockwise   anticlockwise

2 Use a red pencil to mark the direction on each circle.

a half turn
clockwise

a quarter turn
anticlockwise

a whole turn
clockwise

a whole turn
anticlockwise

a half turn
anticlockwise

a quarter turn
clockwise

3 Draw the objects on the tables.

- If the girl makes $\frac{1}{4}$ of a turn clockwise she faces the crayons.

- If the girl makes a whole turn she faces the plants.

- If the girl makes $\frac{1}{4}$ of a turn anticlockwise she faces the paints and brushes.

Date: _____

Geometry and Measure

Lesson 1: **Using tally charts to collect data**

• Use a tally chart to collect and interpret data

1 Complete the tally chart.

Vehicles

Vehicle	Tally
truck	
car	
motorbike	

2 Use the tally chart to answer the questions.

Vehicles

Vehicle	Tally
truck	ℍℍ ℍℍ ////
car	ℍℍ ℍℍ ℍℍ ℍℍ ///
motorbike	ℍℍ ℍℍ /
taxi	ℍℍ

How many?

a cars [ ]

b trucks [ ]

c taxis [ ]

3 Use the tally chart in **2** to answer the questions.

a How many more trucks than taxis? [ ]

b How many more cars than motorbikes? [ ]

c What does this tally chart tell us? _____

_____

Date: _____ 🙂 😐 ☹

Statistics and Probability

Lesson 2: **Pictograms and block graphs**

• Use a pictogram or block graph to present data

You will need
• coloured pencil

1 Complete the pictogram for the number of animals in the zoo.

Animals in the zoo

Animal	Number
elephant	2
tiger	4
monkey	8

Animals in the zoo

Animal	Number
elephant	
tiger	
monkey	

2 Draw a pictogram for the number of milkshakes sold.

Milkshakes sold

Milkshake	Number
banana	9
strawberry	2
chocolate	5
mango	10

Milkshakes sold

Milkshake	Number

Statistics and Probability

3 Use the data in the table in **2** to complete the block graph for the number of milkshakes sold.

Milkshakes sold

Number of milkshakes

banana   strawberry   chocolate   mango

What do the pictogram and the block graph tell us?

_____

4 Use the data in the pictogram to answer the questions.

Favourite sports

Sport	Number
cycling	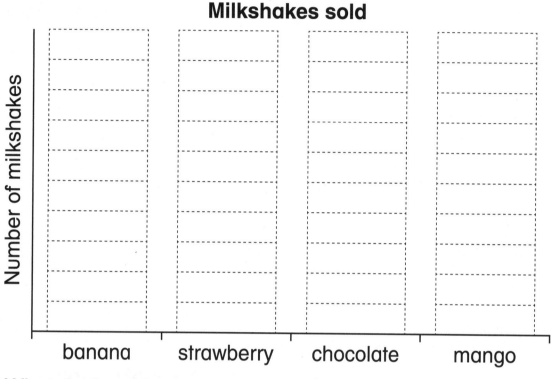
basketball	
tennis	
swimming	

a How many people like cycling? ☐

b How many more people like basketball than tennis? ☐

c How many people were surveyed altogether? ☐

d Does the pictogram tell us what everyone's favourite milkshake flavour is?

| yes | no |

e Why/why not? _____

Date: _____

Lesson 3: **Venn diagrams**

• Read and create Venn diagrams with two sorting rules

1 Draw each shape in the correct part of the Venn diagram.

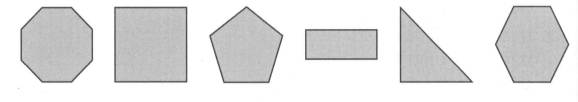

Shapes

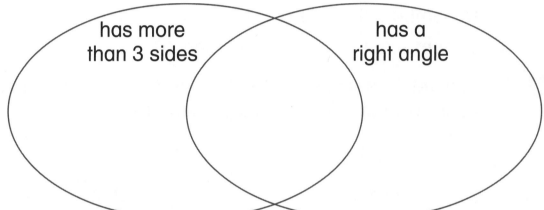

has more
than 3 sides

has a
right angle

2 Work out the two sorting rules, then write the labels.

Beetles

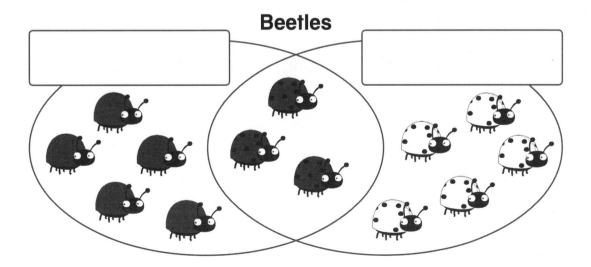

Statistics and Probability

Statistics and Probability

3 Write these numbers on the Venn diagram:
1, 2, 3, 4, 5, 6, 7, 8, 9, 10

Numbers

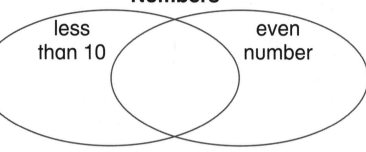

What does the central part of the Venn diagram show us? Draw a ring around the answer.

odd numbers that are less than 10

even numbers that are less than 10

even numbers that are greater than 10

4 Question eight people in your class and write their names on the Venn diagram. Give the diagram a title.

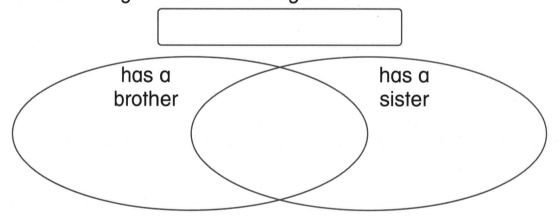

Draw a ring around the things that the Venn diagram does **not** show us.

people who have a brother and a sister

people who only have a brother or brothers

people who wish they had a brother

people who do not like their sister

people who only have a sister or sisters

people who have lots of brothers and sisters

Date: _____

Lesson 4: **Carroll diagrams**

• Read and create Carroll diagrams with two sorting rules

You will need
• coloured pencils

1 Draw at least two shapes in each part of the Carroll diagram.

Shapes

	squares	not squares
blue		
not blue		

2 Find out the ages of at least eight learners in your class and write their names in the Carroll diagram.

Our class

	aged 7	not aged 7
boy		
not a boy		

Statistics and Probability

111

Statistics and Probability

3 Use the Carroll diagram to answer the questions.

Girls who can swim and ride a bike

	can ride a bike	cannot ride a bike
can swim	Rani, Mariam	Molly, Jess, Amira
cannot swim	Lily, Lin, Ife	Daima, Emmy

a Can Ife ride a bike?

b How many girls can swim?

c How many girls can ride a bike, but cannot swim?

d Can the Carroll diagram tell us which girls like swimming?

| yes | no |

Why/why not? _____

4 **a** Write the headings to make this Carroll diagram true.

b In the empty box, draw a shape that follows the rules.

Shapes

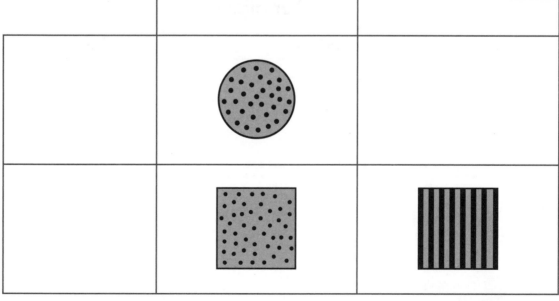

Date: _____

Lesson 1: **Identifying patterns**

• Recognise and describe regular and random patterns

1 Are the patterns regular or random? Draw a ring around the answer.

a

regular   random

b

regular   random

2 Look at these five patterns. Cross out the random pattern. Continue the other four patterns.

a 1, 5, 6, 1, 5, 6, 1, 5, ☐ , ☐

b

c

d A, D, D, H, A, D, D, H, A, ☐ , ☐

e

3 Are the days of the week a regular or random pattern?

regular   random

Why? _____

4 Is what you have for breakfast on each day a regular or random pattern?

regular   random

Why? _____

Date: _____

Statistics and Probability

Lesson 2: **Chance**

Statistics and Probability

You will need
• red and blue coloured pencils

• Look at patterns in chance experiments

1 Tick the only shape that you could pick from this bag.

square   circle

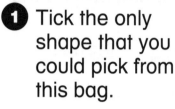

2 Tick the shapes that you could pick from this bag.

circle   square
triangle   star

3 I want a chance of picking either a red ball or a blue ball. Colour the balls to make this possible.

4 Team A picked the shapes from the bag in this order:

Team B picked the shapes from the bag in this order:

Why do you think the results are different?

_____

5 There are 10 jellybeans and 2 toffees in a bag. 12 children pick a sweet. Will more children get jellybeans than toffees? How do you know?

| yes | no |

_____

_____

_____

How could you make it more likely to pick a toffee?

_____

_____

Date: _____

Lesson 3: **Investigating chance**

- Investigate chance and record the results

1 Complete the tally chart.

2 Was picking a red ball or a blue ball more common?

red ball	blue ball

	Tally	Total
red ball	‖‖ ‖‖ ‖‖‖‖	
blue ball	‖‖‖‖	

3 A class investigated the outcome of spinning a two-colour spinner.

a Complete the tally chart.

	Tally	Total
pink		12
green		18

b Do you think the results will be the same or different if they do the investigation again?

same	different	Why? _____

_____

4 A class investigated the chance of getting heads when a coin is flipped. Is the result in this tally chart possible?

	Tally	Total
heads	‖‖‖‖ ‖‖‖‖ ‖‖‖‖ ‖‖‖‖ ‖‖‖‖	24
tails	‖‖‖‖	5

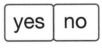

Why? _____

_____

Date: _____

Statistics and Probability

Statistics and Probability

Lesson 4: **Presenting and describing data**

• Present and describe data on a chart or graph

You will need
• coloured pencil

There were 16 cookies in a tin. 10 were chocolate and 6 were plain. 12 people took a cookie without looking inside the tin. Here are the results.

Cookies

Cookie	Number
chocolate	
plain	

1 How many people got a chocolate cookie? ▢

2 Which type of cookie did most people get? ▢

3 How many more people could pick a chocolate cookie from the tin? ▢

4 If you took a cookie from the tin now, do you think you would get a chocolate cookie?

| yes | no |

Why? _____

_____

5 Why do you think more people picked a chocolate cookie than a plain cookie?

_____

_____

Statistics and Probability

6 What would you have to change about the cookies if you wanted to repeat the investigation and make sure that everybody got a plain cookie?

_____

_____

7 Draw a block graph to show what you think would happen if there were 15 chocolate cookies and 1 plain cookie in the tin and 12 people took a cookie.

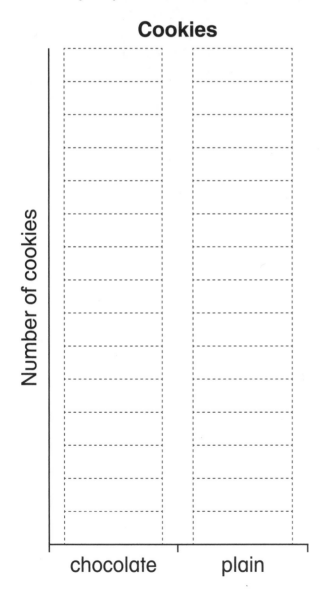

Cookies

Number of cookies

chocolate   plain

Date: _____

Acknowledgements

Photo acknowledgements

Every effort has been made to trace copyright holders. Any omission will be rectified at the first opportunity.

p9t Tartila/Shutterstock; p9b Omeris/Shutterstock; p14t Berkah Icon/Shutterstock; p38 Ponysaurus/Shutterstock; p38tr Sudowoodo/Shutterstock; p41c Ortis/Shutterstock; p41b Wu Hsiung/Shutterstock; p42b Olkita/Shutterstock; p46cb Archivector/Shutterstock; p80c Szefei/Shutterstock; p80b Attaphong/Shutterstock; p81tl Alexandr III/Shutterstock; p81tcl MPFphotography/Shutterstock; p81tc Olga Kovalska/Shutterstock; p81tcr Surot Kumthong/Shutterstock; p81tr 25krunya/Shutterstock; p81b Szefei/Shutterstock; p85tl Alona_S/Shutterstock; p85tc Yulistrator/Shutterstock; p85tr Selcuksevindik/Shutterstock; p85cl GalapagosPhoto/Shutterstock; p85c Virtu studio/Shutterstock; p85cr Gomolach/Shutterstock; p93t Black creator 24/Shutterstock; p93c Tanachai Chaisri/Shutterstock; p93b Hayati Kayhan/Shutterstock; p113bl Mallinka1/Shutterstock; p113bc MOSAIC/Shutterstock; p113br Anest/Shutterstock; p116t Nattika/Shutterstock; p116b Drebha/Shutterstock; p117 Milart/Shutterstock.